Nurturing Food Justice

Food, Health, and the Environment

Series Editors: Robert Gottlieb, Henry R. Luce Professor of Urban and Environmental Policy, Occidental College

Nevin Cohen, Associate Professor, City University of New York (CUNY) Graduate School of Public Health

Keith Douglass Warner, *Agroecology in Action: Extending Alternative Agriculture through Social Networks*

Christopher M. Bacon, V. Ernesto Méndez, Stephen R. Gliessman, David Goodman, and Jonathan A. Fox, eds., *Confronting the Coffee Crisis: Fair Trade, Sustainable Livelihoods and Ecosystems in Mexico and Central America*

Thomas A. Lyson, G. W. Stevenson, and Rick Welsh, eds., *Food and the Mid-Level Farm: Renewing an Agriculture of the Middle*

Jennifer Clapp and Doris Fuchs, eds., *Corporate Power in Global Agrifood Governance*

Robert Gottlieb and Anupama Joshi, *Food Justice*

Jill Lindsey Harrison, *Pesticide Drift and the Pursuit of Environmental Justice*

Alison Alkon and Julian Agyeman, eds., *Cultivating Food Justice: Race, Class, and Sustainability*

Abby Kinchy, *Seeds, Science, and Struggle: The Global Politics of Transgenic Crops*

Vaclav Smil and Kazuhiko Kobayashi, *Japan's Dietary Transition and Its Impacts*

Sally K. Fairfax, Louise Nelson Dyble, Greig Tor Guthey, Lauren Gwin, Monica Moore, and Jennifer Sokolove, *California Cuisine and Just Food*

Brian K. Obach, *Organic Struggle: The Movement for Sustainable Agriculture in the U.S.*

Andrew Fisher, *Big Hunger: The Unholy Alliance between Corporate America and Anti-Hunger Groups*

Julian Agyeman, Caitlin Matthews, and Hannah Sobel, eds., *Food Trucks, Cultural Identity, and Social Justice: From Loncheras to Lobsta Love*

Sheldon Krimsky, *GMOs Decoded: A Skeptic's View of Genetically Modified Foods*

Rebecca de Souza, *Feeding the Other: Whiteness, Privilege, and Neoliberal Stigma in Food Pantries*

Bill Winders and Elizabeth Ransom, eds., *Global Meat: The Social and Environmental Consequences of the Expanding Meat Industry*

Laura-Anne Minkoff Zern, *The New American Farmer: Immigration, Race, and the Struggle for Sustainability*

Julian Agyeman and Sydney Giacalone, eds., *The Immigrant-Food Nexus: Food Systems, Immigration Policy, and Immigrant Foodways in North America*

Benjamin R. Cohen, Michael S. Kideckel, and Anna Zeide, eds., *Acquired Tastes: Stories about the Origins of Modern Food*

Karine E. Peschard, *Seed Activism: Patent Politics and Litigation in the Global South*

Jennifer Gaddis and Sarah Robert, eds., *Transforming School Food Politics around the World*

Mariana Chilton, *The Truth about Hunger in America: And the Resolve to Heal It through Justice, Solidarity, and Love*

Jia-Chen Fu, Michelle T. King, and Jakob A. Klein, eds., *Modern Chinese Foodways*

Adrian Hearn, *Food System Intermediaries: Bonding and Bridging in China, Latin America, and Australia*

Alison Hope Alkon and Julian Agyeman, eds., *Nurturing Food Justice: Expansive and Intersectional Visions*

Nurturing Food Justice

Expansive and Intersectional Visions

edited by Alison Hope Alkon and Julian Agyeman

The MIT Press
Cambridge, Massachusetts
London, England

The MIT Press
Massachusetts Institute of Technology
77 Massachusetts Avenue, Cambridge, MA 02139
mitpress.mit.edu

The MIT Press would like to thank the anonymous peer reviewers who provided comments on drafts of this book. The generous work of academic experts is essential for establishing the authority and quality of our publications. We acknowledge with gratitude the contributions of these otherwise uncredited readers.

This book was set in Stone Serif and Stone Sans by Westchester Publishing Services. Printed and bound in the United States of America.

Library of Congress Cataloging-in-Publication Data

Names: Alkon, Alison Hope, editor. | Agyeman, Julian, editor.
Title: Nurturing food justice : expansive and intersectional visions / edited by Alison Hope Alkon and Julian Agyeman
Other titles: Food, health, and the environment.
Description: Cambridge, Massachusetts : The MIT Press, 2025. | Series: Food, health, and the environment | Includes bibliographical references and index.
Identifiers: LCCN 2025010960 (print) | LCCN 2025010961 (ebook) | ISBN 9780262553698 (paperback) | ISBN 9780262385343 (epub) | ISBN 9780262385350 (pdf)
Subjects: LCSH: Food supply—Moral and ethical aspects—United States. | Food industry and trade—Moral and ethical aspects—United States. | Food sovereignty.
Classification: LCC HD9000.5 .N88 2025 (print) | LCC HD9000.5 (ebook)
LC record available at https://lccn.loc.gov/2025010960
LC ebook record available at https://lccn.loc.gov/2025010961

EU Authorised Representative: Easy Access System Europe, Mustamäe tee 50, 10621 Tallinn, Estonia | Email: gpsr.requests@easproject.com

Contents

Series Foreword

Nurturing Food Justice: Expansive and Intersectional Visions is the twenty-fifth book in the Food, Health, and the Environment series. The series explores the global and local dimensions of food systems and the issues of access; social, environmental, and food justice; and community well-being. Books in the series focus on how and where food is grown, manufactured, distributed, sold, and consumed. They address questions of power and control, social movements and organizing strategies, and the health, environmental, social, and economic factors embedded in food-system choices and outcomes. As this book demonstrates, the focus is not only on food security and well-being but also on economic, political, and cultural factors and regional, state, national, and international policy decisions. Food, Health, and the Environment books therefore provide a window into the public debates, alternative and existing discourses, and multidisciplinary perspectives that have made food systems and their connections to health and the environment critically important subjects of study and for social and policy change.

Robert Gottlieb, Occidental College

Nevin Cohen, City University of New York (CUNY)
Graduate School of Public Health

Touching Dirt, Together: On Nurturing in and Beyond the Political Moment

At a recent conference, we (Alison and Julian) had the opportunity to be in conversation with many of the authors whose words you'll read in this book. After each described their contribution, we asked them to think about how their work might speak to the challenges of our political moment. The rapid, flood-the-zone onslaught of executive orders and other policies can feel dizzying, and we don't know what this world will look like by the time you hold this book in your hands. Still, we felt a deep dissonance between the expansive, intersectional stories of more just and sustainable food futures that we offer you here and the amplified cruelty of this administration, cruelty that will make so many of us, in the words of Ruth Wilson Gilmore, disproportionately vulnerable to premature death. Our conversation helped us to see the foundations of the bridges we can build, so we thought it fitting to share.

At the conference, Lex Wiley started us off with a response rooted in the practices of historical and contemporary fugitivity that animate their chapter, reminding us that Black and Indigenous peoples have long experienced catastrophe. Ike Leslie echoed this, arguing that the Democrats may have just discovered fascism, but many communities, including the LGBTQ+ farmers they write about, have been subjected to state persecution for a very long time.

But woven into these dark provocations was a return to the capacious, hopeful tone you'll find throughout this volume. Charlotte Glennie Roberts invoked the notion of "interstitial spaces," those cracks in the concrete in which new seeds can sprout and grow. Belinda Ramírez called in the concept of "maroonage" to bring in "examples of flourishing beyond the state" that could offer new understandings of "what it means to be human." Ash

McLeod, whose chapter examines how a Native-owned nonprofit and restaurant understand decolonial food, reminded us that the ongoing catastrophe that Lex referred to has always been paired with resistance, declaring that Native communities are "laying the groundwork for new folks to take up this ongoing project and make the world better for the next Seven Generations." And Lex's coauthor, Akilah Chatman, invoked the grounding potential of land-based work in moments of upheaval, reminding us all to "touch dirt, together."

Each of these writers was calling on the prefigurative power of the food justice movement to help us imagine and practice new worlds even as the hegemonic one leans into new and intensified forms of violence. The movements and communities we work with and write about have taught us that food can be a powerful tool for organizing, for bringing people together across differences, and for building power. In the days after the 2025 inauguration, we often heard the phrase, "The way we get through this is together." Food, and food movements, have a lot to offer as we cultivate this togetherness.

And yet there are tensions too. Laura-Anne Minkoff-Zern drew on her chapter and new book (both cowritten with Teresa Mares) to remind us that calls for community care and mutual aid must be tempered by the realization that these voluntary efforts can add to women's uncompensated workload and can draw attention from the underpaid and undervalued food system labor performed disproportionately by precarious immigrant and other BIPOC workers. And food scholars know well that the community-based work of feeding one another too often fails to fill the gaps left by a state that is now even less committed to meeting basic needs. Indeed, the most successful community food efforts, like the Black Panther Party's free breakfast program, don't just replace state efforts but force the state to act.

We will need both these sharply critical insights and expansive radical imaginaries in order to forge solidarities in the times ahead. We will need to stand with farm workers and student activists targeted for deportation and extraordinary rendition; with communities in the US and globally experiencing intensified hunger as food aid is decimated; with farmers, researchers, and government workers who have been abandoned by the USDA, the FDA, and other federal agencies; and with whoever ends up in the crosshairs of cruelty next.

The challenge, as always, is to hold the big picture and the current reality, to live our lives in a way that brings more just and sustainable futures into being while also standing strong against the contemporary moment. It is hard to think creatively amid the chaos, but this book offers many potential avenues for doing so: Can the retrenchment of state-funded basic needs offer an opportunity to build on the work of the food banks described in chapter 10 to reimagine the right to food? Can the Make America Healthy Again contingency create an opening to think about the policies and practices that actually threaten our collective and public health or to deconstruct commonsense understandings of what health even is, which is the question taken up in chapter 9? Can our contemporary threats to land and labor offer new insights that build on the movement work described in chapter 1 to better understand the nature of contemporary racial capitalism and to find new pressure points from which to respond? In these pages, you'll find strategies for fugitive survivance, for engaging with institutions, and for centering essential workers once again. We hope it will resonate with you in ways we have not yet imagined, in our current political moment and the world we build together in its wake.

In community, Alison and Julian, with love and gratitude to our session participants (alphabetically)—Akilah Chatman, Greig Guthey, Ike Leslie, Ash McLeod, Laura-Anne Minkoff-Zern, Belinda C. Ramírez, Charlotte Glennie Roberts, Alana Haynes Stein, and Lex Wiley—and to all of our chapter authors, and to the activists and community partners who envision and create just and sustainable food futures

Introduction: Nurturing Just and Sustainable Food Futures

Alison Hope Alkon and Julian Agyeman

When we published *Cultivating Food Justice* in 2011, we began by asking readers to "picture a field of corn stretching out into the horizon." We described its genetic homogeneity and the mechanization and economic centralization that both produce and are reproduced by it. Though social movements to reform and transform the food system condemn this sort of industrial agriculture as socially and ecologically destructive, we used this metaphor to suggest that these movements themselves were a sort of cultural monoculture. Not only had white activists become the most prominent voices for alternative food systems, but the narratives and strategies they brought to this work were infused with cultural norms and values rooted in an agrarian imaginary that resonated most deeply with white people, inviting them in while subtly pushing Black, Indigenous, and people of color (BIPOC) communities away.

Today we begin by asking you to picture a garden. The soil is dark, enriched by diverse agroecological knowledges and techniques. In one bed you might find the "three sisters," as cultivated by Indigenous peoples across Turtle Island. Several varieties of nitrogen-fixing pole beans are planted in mounds together with maize, which provides a trellis as they grow. On the earth below, squash leaves offer shade to the soil. In other comingled beds, there are yams, collards, and okra, grown from seeds first brought to the Americas by enslaved Africans before becoming a part of the American diet (Keeve 2020). There are bittermelon and long beans cultivated by Southeast Asian refugees, purple potatoes from the Peruvian highlands, and hundreds of varieties of chiles and tomatoes first domesticated in Mexico. Next to the tomatoes, *cempasuchil*, or marigolds, represent a connection to ancestors, especially during Dia de los Muertos, while their use in Hindu ceremonies reminds us that plants can connect cultural traditions across the globe

(Akhtar 2023)[1]. *Cempasuchil* can also provide habitat for beneficial insects and organic pest management techniques as an alternative to soil fumigation.

This garden is a polyculture, a diversity of plants and practices that embodies the traditions, knowledges, practices, and stories that food justice activists draw on to create new forms of sustenance. If "freedom is a place," as abolitionist geographer Ruth Wilson Gilmore compellingly argues (2017, 227), it may well look like this garden.

This polyculture is nourished by a movement for food justice and by what Robin D. G. Kelly (2022) calls the "freedom dreaming" of activists cultivating visions of the world they seek to build. Practices of growing and sharing food are often associated with what bell hooks describes as "communities of care" (2009, 292; Kohl and McCutcheon 2015), and it is this throughline (brown and Brown 2022) that we meant to invoke as we chose our title. In the wake of the COVID-19 pandemic, care has also become a more explicit element of radical social theory (Moulton et al. 2021) and foregrounds long-overdue feminist approaches to social movement work (Hayes and Kaba 2022; Hobart and Kneese 2020; Allen and Sachs 2007; Parker 2019). And as we and the majority of our contributors work in academia, we also highlight the importance of care as an essential lens through which to approach our own endeavors within and beyond our institutions (Anguelovski 2020). Cultures of care, according to environmental justice scholar Phaedra Pezzulo, are constituted "through shared values and experiences of what it means to collectively imagine and manifest more viable futures [and to] navigate the traumatic times we are living in" (2023, 11). Care is both an analytic framework and a way of being as a community of scholar-activists. In this spirit, we offer this volume as a way to nurture the food justice movement as it grows.

And yet, despite the way food is so often linked to care, there is nothing inherently nurturing about food, and nurturing, like nourishment, can have multiple and contextually dependent meanings. Food is necessary for our survival and is a way to celebrate our collective flourishing, but it also contributes to disease, inequality, and health disparities (Gálvez et al. 2020). Food can bring communities together (Julier 2013, McCutcheon et al. 2023) but can also instill divisions based on power and prestige (Parasecoli 2022). Food can be a source of economic development, cultural expression, and sovereignty (Morrison 2011; Agyeman and Giacalone 2020; Ray 2016) and also a way to shame, surveil, and regulate (Williams-Forson 2023; Reese and Sbicca 2022; Dickinson 2019; DeSouza 2021). It is a multi-trillion-dollar industry subsidized not only by the US government but by the often invisible (and invisibilized), essential, dangerous, and un/underpaid labor of the often-overlapping categories of immigrants, people of color, and women (Carney 2015; Minkoff-Zern and Mares 2025; Brenton et al. 2019; McCutcheon et al. 2013).

The food system acts as a racial project, establishing and maintaining racial hierarchies and subjectivities (Omi and Winant 2014; Sbicca and Meyers 2017). It is a site where racialized political ecologies become embodied (Jones 2019; Hayes-Conroy and Hayes Conroy 2010) as an assemblage of slow violences (Nixon 2011) that contribute to our "group-differentiated vulnerabilities to premature death" (Gilmore 2008, 261). Activists have long been working to transform this system into one that is sustainable, inclusive, and just. But even within food movements, white-led organizations receive disproportionate funding and visibility (Reynolds and Cohen 2016; McClintock and Simpson 2017) and often elide the racialized agricultural histories and geographies through which communities of color relate to land, food, and agriculture (Ramírez 2015; Alkon 2012; Guthman 2008). Moreover, capitalist real estate imperatives can position community gardens, restaurants, and farmers markets as market signals for investment opportunities, resulting in the further dispossession of marginalized communities (Alkon, Kato, and Sbicca 2020), while well-funded entrepreneurs appropriate BIPOC, working class and immigrant foodways (Agyeman and Giacalone 2020). The food justice movement does not ignore these divisions, hierarchies, and oppressions in a blithe celebration of gardens and cuisine but actively seeks to transform them.

The Praxes of Food Justice

Many activists and scholars have attempted to characterize this complex movement, even as it consistently and collectively reflects and evolves. In the earliest scholarly collection on food justice, Gottlieb and Joshi (2010) highlighted the fair allocation of the risks and benefits produced by the production, distribution, and consumption of food. This definition foregrounds food justice's taproot in struggles for environmental justice, which remains essential as both fields move beyond earlier, distribution-based appeals to interrogate the ways that race, space, and nature are coproduced (Pulido and De Lara 2018; Meehan et al. 2023; Van Sant et al. 2020; Pellow 2016). Weaving together critical race theories with political ecologies of food systems, food justice researchers answer Pulido's challenge to go beyond documenting disparities and "theorize environmental racism as a constituent element of racial capitalism," (Pulido 2017, 524; Robinson 1983) as well as settler colonialism (Whyte 2018) and carceral (agro)ecologies (Sbicca and Reese 2022).

More than 10 years ago, in *Cultivating Food Justice* (Alkon and Agyeman 2011), we selected the definition used by the veteran nonprofit organization Just Food because we valued its right-to-food approach and its emphasis on "care for the well-being of the land [and] workers." Land and labor remain essential components of food systems and have been the subjects of important writings in their own right (Williams and Holt-Giménez 2017; White 2019; Gaddis 2019; Minkoff-Zern and Mares 2025; Sbicca 2018; Guthman 2004) but are rarely included among food justice's defining features. Chapters 4, 8, 16, and 17 of this volume continue the essential work of making these concerns central.

Food justice is often depicted as one node in an interconnected web of food movements and is distinguished by its focus on institutional racism (Motta 2021). In this vein, activist-farmer Rasheed Hislop defined the movement as "the struggle against racism, exploitation, and oppression taking place within the food system that addresses inequality's root cause both within and beyond the food chain" (2015, 24; see also Morales 2011 and Penniman 2018). Anthropologists Hannah Garth and Ashanté Reese continue to center race, and anti-Blackness in particular, in their groundbreaking (or groundhealing) book *Black Food Matters: Racial Justice in the Wake of Food Justice*. They name food system inequities as constituent of the afterlife of slavery and the ongoing devaluation of Black life (Hartman 2007). In response, food justice stands as a "theoretical, methodological, and aspirational framework for re-envisioning a world in which access to healthy, affordable food is not a dream but a reality for all" (Garth and Reese 2020, 1).

The everyday work of the food justice movement is in constant dialogue with these varied understandings of what it is and can be, creating an array of dynamic and theoretically informed praxes that craft analysis and action at the nexus of food systems and inequalities. As an academic subfield, food justice research values scholar-activism, community-based studies, and deep engagement with social movement work. Indeed, Kristin Reynolds, Daniel Block, and Katherine Bradley (2018), three scholars whose research has been exemplary in this regard, question how well the binary of scholar and activist maps onto the food justice literature, given that those working primarily in university settings participate in social movements while those working for food justice organizations engage in research and writing for academic and general audiences (see also Levkoe et al. 2020). Indeed, five of our twenty chapters include activist coauthors. This is not to argue that institutional privilege does not matter; indeed, the antiracist, decolonial, and feminist approaches that guide the chapters in this volume demand that we sit with our multiple, intersecting and often complex, positionalities. And yet, despite the institutional barriers to this type of research, we know that cross-pollination between scholarship and activism greatly enriches the field.

From our own standpoints as academics who have worked with, written with, and supported community-based organizations over the past twenty years, we attempt to offer a capacious synthesis of what food justice is and can be that encompasses aspects of kindred efforts such as food sovereignty and agroecology. Our deep engagement with food justice activism and scholarship informs a vision that coheres around three essential nodes. First, scholars have documented the ways that various past and present policies and projects have limited *food access* in communities that experience disinvestment, overpolicing, and what Gilmore (2022) calls "organized abandonment" (McClintock 2011, 2018; Alkon 2012; Kolavalli 2023; Meyers 2022). The roots of this uneven food access lie in the initial and ongoing theft of Indigenous land on which settler-colonial nations were founded, the kidnapping of Black Africans to provide agricultural labor, and the afterlives of these horrors, including racist land-use policies, economic incentives, uneven development, and the institutional racism experienced by Black, Indigenous, Mexican, Asian and women farmers (Garth and Reese 2020; McClintock 2011; Jett 2021; Figueroa and Penniman 2020; Minkoff-Zern et al. 2011). Contemporary BIPOC neighborhoods in which people have withstood and made lives within these dynamics have often been labeled "food deserts" for their lack of grocery stores, a term that continues to hold significant currency in policy and planning circles (Zurawski 2023). Activists, however, compellingly argue that this metaphor is a form of what Unangax̂ scholar Eve Tuck (2009) calls "deficit framing" and that food apartheid better reveals the racial and sociopolitical, rather than natural roots of these unequal built environments (Washington 2018; Reese 2019; Zurawski 2023). Chapters 6 and 7 of this volume offer historical and embodied lenses that contextualize food apartheid, expanding the field's ability to conceptually link food, inequalities, and urban landscapes (Heynen, Kurtz, and Trauger 2012).

In response to these conditions, many activists work to cultivate food and/or to support BIPOC farmers in order to supply culturally relevant food to communities experiencing food apartheid. This approach to food justice positions it as a "redress discourse" aimed at responding to past and ongoing food-related violence (Best and Hartman 2017; see also Tuck and Yang 2017). Researchers have offered detailed accounts of the visions and practices that inform food justice organizations' farmers markets (Alkon 2012; Meyers 2022), community gardens (White 2019; Reynolds and Cohen 2016; chapters 4 and 8, this volume), and cooperative and community-owned grocery stores and restaurants (Figueroa and Alkon 2017; Cooke 2022; chapter 3 this volume). Through these projects, community-based food justice initiatives offer not only food but a celebration of ancestral food traditions and an opportunity to develop practices of care and healing (Gilbert and Williams 2020; Penniman 2018; Cupchoy 2023; Coté 2022; Cadieux and Slocum 2015; chapters 5, 10, 15, and 16 this volume). Food justice thus becomes a set of land-based practices that can nurture the bodies and spirits of communities on the front lines of racial capitalism and settler colonialism.

Food justice scholars have also looked beyond the work of community-based organizations to better understand the everyday provisioning struggles of marginalized people who are not, or not yet, served by these activist projects (Carney 2015; Brenton et al. 2019; Ewoodzie 2021). In seeking to bridge the gap between what activists cultivate and broader community needs, *Nurturing Food Justice* includes efforts to support and shift institutions that reach many people experiencing poverty where they are, including food banks (chapter 11) and street food vendors (chapter 12). In contrast with nutrition education, which locates the root of food disparities in what is presumed to be the deficient knowledge of marginalized people (Guthman 2008; Minkoff-Zern and Carney 2015; chapter 9, this volume), institution-shifting approaches such as these push back against the structural constraints that make it so difficult for BIPOC farmers to make a living and for minoritized and disinvested communities to access their essential foods.

Second, while increasing food access is necessary to the food justice movement, establishing access to land and community-based stewardship for food and agricultural systems is equally important. In this regard, food justice draws upon and amplifies the global peasant-led movement for *food sovereignty*, especially the work of La Via Campesina (Wittman and Desmarais 2010; Borras and Franco 2023). Indigenous scholars and activists within North America position food sovereignty within struggles for decolonization, self-determination, and care for the land (Coté 2022; Whyte 2018; Morrison 2011; chapter 5, this volume), though not all Indigenous communities describe the resurgent food practices through which they reclaim both their kinship and authority in this way (Daigle 2019). In the US, predominantly white communities have invoked food sovereignty in their support for local producers (Trauger 2017), though Fairbairn (2012) warns that, absent an analysis of social injustice, food sovereignty risks becoming a synonym for local control. In contrast, food justice scholarship has long placed Black, Latinx, and Indigenous community autonomy in the context of racial and settler colonial oppressions (McCutcheon 2011; Mares and Peña 2011; Norgaard, Reed, and Van Horn 2011; chapters 4 and 8, this volume) and the settler ecologies through which white colonial powers have claimed and reshaped land for their own ends (Dillon 2021, 2024).

Most fundamentally, the food justice movement puts forward a vision in which food systems rooted in food access and food sovereignty become a means to *transform socioecological relationships* toward just sustainabilities (Agyeman 2013; Agyeman et al/ 2003). Sustainability refers to a regard for our more than human natural systems and encompasses processes that are necessary to avoid, mitigate, and repair ecological disasters. In practice, however, the ideal of sustainability is often co-opted by the eco-modernist promotion of efficiencies and green technologies within capitalism. In contrast, a just sustainability framework also recognizes "the inseparability of environmental quality and human equality" (Agyeman 2008, 751). In this context, food justice is about more than just food. It becomes a means "to determine and create sustainable worlds free of anti-Blackness and its accompanying oppressions" (Garth and Reese, 6), a strategy for creating emancipatory power (Smith 2019), an argument for land-based reparations for Black and Indigenous communities (Penniman 2018; Gilbert and Williams 2020; Jernigan 2021; chapter 19, this volume), and a standpoint from which to "imagine more-than-capitalist food economies" (Morrow et al. 2023; chapter 1, this volume). Food justice is an essential part of decolonial struggles for Indigenous sovereignty (Hoover and Minhesuah 2019; Whyte 2018; chapter 5, this volume) and can inform food worker campaigns that are revitalizing worker organizing (Minkoff-Zern and Mares forthcoming). Food justice is integral to the transformational policy platform put forth by the Movement 4 Black Lives, the Green New Deal, the Red New Deal, and the Alliance for Climate Justice's Just Transition Framework. And it can nurture abolitionist movements that share its "commitments to build a radically different society" (Reese and Sbicca 2022, 3; chapter 20, this volume; Heynen and Ybarra 2020; Thompson 2023; Montenegro de Wit 2021). Taken together, food justice has much to contribute to what Mdewakanton Dakota/Diné activist Dallas Goldtooth calls "this movement of movements," struggles that seek to enact just and sustainable futures beyond racial capitalism, settler colonialism, heteropatriarchy and climate collapse (quoted in Sze 2020, 21).

The Stories of Food Justice

To weave together these definitions and praxes of food justice, and to locate the field within critical social theory, is to offer a story of what the movement is and can become. Sylvia Wynter describes humans as a storytelling species; being human is not only about biological survival but about the narratives that underlie our ways of seeing and being (McKittrick 2015, 2021). Building on this work, Pavithra Vasudevan and her coauthors position the "White, cis male, bourgeois and propertied figure of the human . . . as a fiction of modernity," and look to Black, Indigenous, and third-world women's storytelling, including Wynter, "to reimagine what it means to be human on earth" (2023, 1728). Indigenous food geographies speak of storytelling as both "theory and methodology . . . an act of resistance and a form of decolonization [that] help to shape who we are" (Coté 2022, 9). For example, chapter 5 author Charlotte Coté (Tseshaht/Nuu-chah-nulth) invokes Robin Wall Kimmerer's (2013) *Braiding Sweetgrass* to demonstrate story as a means to convey traditional ecological knowledge while theorizing colonization and ecological catastrophe as intertwined. Kimmerer too invokes the power of narrative, writing that "we can't meaningfully proceed with healing, with restoration, without 're-story-ation." The field of Latinx geographies also builds on cultural storytelling practices like *cuentos* and *testimonios*, as well as the canonical work of feminist queer theorist-poets Gloria Anzaldúa and Cherrie Moraga (Muñoz and Ybarra 2019; Martinez 2020). Taken together, these storytelling traditions hold up everyday acts of survival and resistance (Woods 2017; Hawthorne and Lewis 2023), showing us how "in the context of planetary apocalypse, storytelling can reanimate scholarship as an anticolonial practice grounded in wonder" (Vasudevan et al. 2023, 1731; McKittrick 2021).

In her book *Counterstory*, Aja Martinez offers a parallel cosmology in which storytelling is a central methodology for critical race theory. Counterstories, she argues, can "expose, analyze, and challenge majoritarian stories of racialized privilege and can help to strengthen traditions of . . . survival, resistance, and justice" (Martinez 2020, 26; see also Dixon 2018). Martinez praises the work of foundational critical race theorists not only for their content but for a poetics that integrates narrated dialogue, fantasy, and autobiographic reflection into their scholarship. Stories of food justice from within the academy increasingly transgress the boundaries of the ivory tower to engage in this sort of creative work. Here, we hold up as examples Leigh Anna Hidalgo's (2015) digital humanities work on street food vendors in Los Angeles, which is in dialogue with chapter 12, Naya Jones's (2019) virtual exhibit *Dying to Eat*, Julian Agyeman's Urban Food Stories web project, Danielle Purifoy's exhibit with the abstract painter Torkwase Dyson (2023), which draws on W. E. B. DuBois's 1906 speculative fantasy novel about the liberatory possibilities of Black agriculture in Alabama, as well as the (2023) zine *Blackecologies*, which is close kin to chapter 4. It is worth noting that many of these projects come from the genre-bending field of Black geographies, which weaves together "the poetic, the material, and the affective" dimensions of Black life (Hawthorne and Lewis 2023, 12). These creative and visionary methods from the margins of acceptable academic discourse invoke and embody the lived realities of communities of color.

In *Cultivating Food Justice*, we framed the chapters that highlighted BIPOC food and farming practices as counterstories that could "disrupt the erasures" embodied by the food movement's dominant narrative (Martinez 2020, 26). For some chapter authors, food justice continues to be an important response to the food movement's racial and heteropatriarchal exclusions, which are taking new forms as explicit white supremacy and homo- and transphobia become more culturally and politically acceptable.

However, additional scholarship increasingly positions food justice as its own story, with its own narrative arcs and lineages in the Black Radical Tradition (Garth and Reese 2020; White 2019; chapter 1, this volume), the Civil Rights Movement (Smith 2023; McCutcheon 2019), and the Black Panther Party (Hassberg 2020; Heynen 2009). Indigenous scholars root their work in precolonial stories and food and farming traditions, rarely mentioning the white-dominated food movement at all (Whyte 2018; Coté 2022; chapter 5, this volume). Many of the chapters in our previous volume, *Cultivating Food Justice,* were in dialogue with critiques of the whiteness of the alternative food movement, though it is worth noting that BIPOC chapter authors Priscilla McCutcheon, Teresa Mares, and Devon Peña presaged this shift toward thinking about food justice in terms of autonomy and fugitivity (2011). In *Nurturing Food Justice,* some authors continue to speak to the "mainstream" food movement and other predominantly white institutions, but most are writing additionally or instead within "the unstoppable force of insurgent thinking, abolition geography, critical race theory, Black and Indigenous geographies, scholar activism, and environmental justice praxis" (Meehan et al. 2023) that nurtures land-based worldbuilding practices. This shift in the food justice literature parallels geographer Audrey Kobayashi's "twin analytical move" of first making whiteness visible and then decentering it so that other stories can be told (2010, 2014).

Storytelling is an essential component of numerous radical BIPOC theoretical traditions (Vasudevan et al. 2023), and each of the chapters in this volume do crucial narrative work. In some cases, this will be obvious to readers, as several chapters use story-based methodologies such as autoethnography (chapters 5, 7, and 19) and polyvocal narrative construction (chapter 14 and 16). But even those chapters deploying more traditional analytic approaches make narrative choices, illuminating some thematic elements rather than others. There is no perceived neutrality in storytelling work; the authors in this volume interpret the lives and worlds of those they study in order to offer stories of what food justice is, is for, and can become.

The Stories We Nurture

The seeds of this book began as a second edition to *Cultivating Food Justice.* We could have written an updated introduction and allowed the original chapters to continue to bear fruit. But the field of food justice research has sprouted so many diverse ways to nurture this movement, and this political moment, so we chose to include new stories, research, and analysis instead. We cast our nets widely to bring together a broad group of forty-eight storytellers, prioritizing BIPOC authors, academic-activist collaborations, early-career faculty, and graduate and undergraduate students. There are omissions here that we need to acknowledge as well, as even our aspirations leave continual room for growth. Most chapters look to community-based organizations and social movements, rather than larger institutions or commercial agriculture. We focused on the US, though there are food justice movements across the globe, and disproportionate attention is given to California.

As editors, we sat for a long time with the challenge of organizing these chapters into a cohesive volume, and we worked within several frameworks before settling on the schema we describe below. We share this process in order to highlight the unruly nature of this scholarship. This book could certainly have been organized differently, and we encourage readers to attend to resonances across the various sections and to seek out work published elsewhere that fills some of these gaps.

We begin with a section called "Dreams" in order to foreground the world-building work that food justice praxis does to envision just and sustainable futures, the inner visions that activists cultivate, and the multiple and intersecting ways that food speaks to individual and collective identities. Antonio Roman-Alcalá offers a thematic overview by describing how the visions of BIPOC food justice activists are rooted in Black radicalism, Indigenous resurgence, and anticapitalist and anarchist strategies. Michaela Hoffelmeyer and Ike Leslie add queer theory to this list of guiding stories, showing us how LGBTQ+ communities counter their exclusion from dominant agrarian narratives and economies through collective practices of queer joy. Ash McLeod and Eleanor Shoreman-Ouimet then describe how workers at an award-winning Indigenous restaurant and associated nonprofit envision and practice the decolonization of their diets. Next, Lex Wiley and Akilah Chatman position Black heritage farming on urban lots as a form of fugitivity surviving in the wake of racist urban design and heightened carcerality, a practice that has roots in the garden plots that helped sustain their enslaved ancestors. Charlotte Coté (Tseshaht/Nuu-chah-nulth) then describes her own ancestral food-gathering practices as a lens through which to understand food justice as part of decolonization and healing. What holds these varied stories together is an inward focus, as communities attempt to orient their gaze toward "survival, subsistence, resistance, and affirmation" (Woods 2017, 27) and conceive of just and sustainable food futures "outside of destruction" (Roane and Hosbey 2019, 1).

And yet, despite the desire for autonomous spaces in which to nurture food justice, activists must continue to engage with and explicitly resist racial capitalism, heteropatriarchy and settler-colonialism. This section, which we call "Confrontations," begins with two very different chapters on uneven urban development. First, Gabriel Valle and Greig Guthey historicize conditions of food apartheid in racist policies including Jim Crow, segregation, and redlining. Justin Sean Meyers then asks whether the experiences of early-stage gentrifiers helped to label their new neighborhoods as food deserts, defining them by what they lack. Next, we follow Belinda Ramírez to San Diego, where they worked with long-time Latinx and Black community gardeners whose access to city-owned land was disrupted by white urban farmers with outsized influence on city government. Lastly, Lucy Aphramor's chapter depicts a confrontation with white supremacy and ableism, not in the city but in the medicalized field of public health nutrition. They argue that we should abolish the colonial construct of "health" achieved through personal responsibility in favor of a politics of care, repair, and collective flourishing. In each of these chapters, food justice activists explicitly work to transform the systems that create unjust food landscapes.

In the wake of the 2020 racial justice uprisings, many institutions responded with stated commitments to incorporate diversity, equity, and inclusion into their missions. This incorporation has been uneven at best and can serve to pacify and diffuse more radical demands. Our third section, "Encounters," offers stories of how activists committed to antiracism and decolonization are working within various institutions with the aim of shifting them, thus extending the reach of the food justice movement. We chose the title "Encounters" in part for its ambiguity, as the work that activists can do within these institutional contexts is continuing to emerge. Alana Haynes Stein, for example, describes how some food banks have begun to expand their programs to incorporate food justice. The next two chapters show how food justice activists can move urban planning efforts toward socioecological justice. Charlotte Glennie Roberts and Catherine Brinkley detail how street food vendors have organized collectively to push the Los Angeles city council to legalize their practices, which increases access to culturally relevant foods and provides opportunities for food-based entrepreneurship in low-income and, especially, immigrant communities. Next, Antwi Akom, Tessa Cruz, and Analena Hassberg describe the power of a digital app called Streetwyze to decolonize data and design by offering a platform through which local community members map their food environments to participate in urban planning. The final three chapters take place within universities. Despite the challenges of what Porter and Wechsler (2018) call academic supremacy, especially in predominantly white institutions, a collective of academics and activists from North Carolina narrate their advocacy for a process that slowed down their timeline, centered the leadership of Black and Indigenous community members, and fostered healing and relationship building. Whitney Barr and Nik Heynen examine how Spelman College's revived victory garden can reimagine what "victory" means while offering Black women students land-based healing from environmental and racial traumas. And, lastly, a collaborative group of authors from the University of Arizona's Future of Food and Social Justice Youth Storytelling Project describe how they developed their own stories in order to counter the enclosure of Tucson's food narrative as a means for gastrodevelopment. In each of these cases, food justice activists expand their praxis beyond food provisioning through encounters with the sometimes-inhospitable contexts of the emergency food system, departments of urban planning, the tech sector, and neoliberal universities.

Our last section, called "Intersections," demonstrates how food justice can reach outward by building bridges with movements and campaigns that are not generally considered central to its work. The first two chapters focus on food labor. Drawing from their recent book, Teresa Mares and Laura-Anne Minkoff-Zern describe three important food worker campaigns in which grocery employees have unionized, farmworkers have achieved state-level demands, and fast-food workers have led the struggle to raise the minimum wage to $15 in cities across the country. Vera Chang then describes an emergent form of labor organizing, worker-driven social responsibility, discussing how farmworkers have transformed exploitative industrial agriculture systems through labor sovereignty—claiming their right to safe workplaces, fair wages, and a voice in supply chain governance. Next, Chellamal Keshavan draws on a Black feminist approach to position "first food equity" for infants and new parents as a form of food justice, and begins to theorize how public health and policy support can become a form of reparation. Maggie Dickinson then explores the space between food justice activism and the Green New Deal and finds that, while there are significant overlaps, it remains necessary to center the voices of communities of color in deciding how this visionary platform should be pursued. Lastly, Joshua Sbicca, Carrie Freshour, Kanav Kathuria, and Sara Black envision abolitionist food justice within and beyond carceral spaces, which can "build life-affirming social and material worlds, which do not require dispossession, displacement, and confinement for their coherence." In some ways, this chapter takes us full circle, as it thinks with the radical traditions that inform food justice freedom dreams while adding an additional layer of what freedom can mean.

There are certainly spaces of overlap between these perspectives and approaches. Whitney Barr and Nik Heynen's chapter about the garden at Spelman College illuminates the healing that the space provides for Black women in a way that is conversant with food justice dreams as well as institutional encounters. Sbicca et al.'s chapter on abolitionist food justice is among the book's most confrontational, in that it describes struggles that take on the ongoing power of racial capitalism embodied by prisons, but it is also an intersection with a kindred movement. We offer these examples to highlight consistencies across variously positioned struggles, which as Maggie Dickinson notes in chapter 19, aligns with adrienne marie brown's (2017) insight about the fractal nature of social movements. In a fractal, the microscale mirrors the macro. Care for the soil becomes care for the communities that the movement seeks to uplift, which becomes care for the new worlds we are building together.

Even all together, these chapters offer only a glimpse of the diversified and fertile fields in which the seeds of food justice have been planted and continue to grow. But we hope they cohere to convey an expansive, dynamic, and intersectional vision of food justice praxes. We are inspired by communities that seek land-based healing from racism, heteropatriarchy, and settler-colonialism in the movement's gardens and renewed food traditions, and we hope these practices can offer respite and even joy as activists continue to confront and resist renewed forms of domination. We are also heartened to see that individuals not directly involved in food cultivation can benefit from the movement's work to increase food access and from its transformative critiques of uneven racialized urban development and unjust labor relations and predatory policing. Most of all, we are uplifted by food justice activists' contributions to the wave of radical uprisings, from Black Lives Matter to Indigenous sovereignty, from climate justice to workers' rights to abolition, and to all of their intersections and spaces in between. In varied ways, each of these movements seeks the transformation of socioecological relations toward just and sustainable futures. We urge readers to pick up your shovels, your voices, and your pens and to find your place in these struggles.

Note

1. With thanks to Sana Javeri Kadri, who taught us about the cultural significance of marigolds.

References

Agyeman, Julian. 2008. "Toward a 'Just' Sustainability." *Continuum* 22 (6): 751–756.

Agyeman, Julian. 2013. *Introducing Just Sustainabilities*. London: Zed Books.

Agyeman, J., R. Bullard, B. Evans, eds. 2003. *Just Sustainabilities: Development in an Unequal World*. Cambridge, MA: MIT Press.

Agyeman, Julian, and Sydney Giacalone. 2020. *The Immigrant Food Nexus*. Cambridge, MA: MIT Press.

Akhtar, Allana. 2023. Marigolds: How a Mexican Flower Bloomed in South Asia. *The Juggernaut*. https://www.thejuggernaut.com/marigold-history-india-mexico-monsoon-wedding.

Alkon, Alison Hope. 2012. *Black White and Green*. Athens, GA: University of Georgia Press.

Alkon, Alison Hope, and Julian Agyeman. 2011. *Cultivating Food Justice*. Cambridge, MA: MIT Press.

Alkon, Alison Hope, Yuki Kato, and Josh Sbicca. 2020. *A Recipe for Gentrification*. New York: NYU Press.

Allen, Patricia, and Carolyn Sachs. 2007. "Women and Food Chains: The Gendered Politics." *International Journal of Sociology of Agriculture and Food* 15 (1): 1–23.

Anguelovski, Isabelle. 2020. "Academia in the Time of COVID-19." *Green Inequalities*. https://www.bcnuej.org/2020/03/31/academia-in-the-time-of-covid-19-our-chance-to-develop-an-ethics-of-care/.

Best, Stephen, and Saidiya Hartman. 2017. "Fugitive Justice." *Representations*. 92(1): 1–15.

Borras, Saturnino M., and Jennifer C. Franco. 2023. *Scholar-Activism and Land Struggles*. Rugby, UK: Practical Action.

Brenton, Joslyn, Sarah Bowen, and Sinikka Elliot. 2019. *Pressure Cooker*. Oxford, UK: Oxford University Press.

brown, adrienne marie, and Autum Brown. 2022. "Care Is the Throughline." *How to Survive the End of the World*. June 9. https://endoftheworldshow.org/search?s=Care+Is+the+Throughline.

Cadieux, Kristen Valentine, and Rachel Slocum. 2015. "What Does It Mean To Do Food Justice." *Journal of Political Ecology* 22 (1): 1–26. https://doi.org/10.2458/v22i1.21076.

Carney, Megan. 2015. *The Unending Hunger*. Berkeley: University of California Press.

Cooke, Christina. 2022. "Co-op Grocery Stores Expand their Mission to Equity and Food Justice." *Civil Eats*. https://civileats.com/2022/09/19/co-op-grocery-stores-expand-their-mission-to-equity-and-food-justice/.

Coté, Charlotte. 2022. *A Drum in One Hand, A Sockeye in the Other*. Seattle: University of Washington Press.

Cupchoy, Lani. 2023. "Healing Our Histories Through Otros Saberes." *Latino Studies* 21: 364–387. https://doi.org/10.1057/s41276-023-00416-4.

Daigle, Michelle. 2019. "Tracing the Terrain of Indigenous Food Sovereignties." *Journal of Peasant Studies* 46 (2): 297–315. https://doi.org/10.1080/03066150.2017.1324423.

DeSouza, Rebecca. 2021. *Feeding the Other*. Cambridge, MA: MIT Press.

Dickinson, Maggie. 2019. *Feeding the Crisis*. Berkeley: University of California Press.

Dillon, Lindsey. 2021. "Civilizing Swamps in California." *Environment and Planning D* 40 (2): 258–275. https://doi.org/10.1177/02637758211026317.

Dillon, Lindsey. 2024. *Toxic City: Redevelopment and Environmental Justice in San Francisco*. Berkeley: University of California Press.

Dyson, Torkwase, and Danielle Purifoy. 2023. "In Conditions of Fresh Water." https://www.torkwasedyson.com/full-project-description Accessed March 31, 2025.

Ewoodzie, Joseph Jr. 2021. *Getting Something to Eat in Jackson*. Princeton, NJ: Princeton University Press.

Fairbairn, Madeleine. 2012. "Framing Transformation: The Counter-Hegemonic Potential of Food Sovereignty in the U.S. Context." *Agriculture and Human Values* 29 (2): 217–230.

Figueroa, Meleiza and Leah Penniman. 2020. "Land Access for Beginning and Disadvantaged Farmers." Green New Deal Policy Series: Food and Agriculture. Chrome-extension://efaidnbmnnnibpcajpcglclefindmkaj/https://www.filesforprogress.org/memos/land_access_for_beginning_disadvantaged_farmers.pdf. Accessed 3/31/2025.

Figueroa, Meleiza, and Alison Hope Alkon. 2017. "Cooperative Social Practices, Self-Determination, and the Struggle for Food Justice in Oakland and Chicago." In *The New Food Activism*, edited by Alison Alkon and Julie Guthman. Berkeley: University of California Press.

Gaddis, Jennifer. 2019. *The Labor of Lunch*. Berkeley: University of California Press.

Gálvez, Alyshia, Megan Carney, and Emily Yates-Doerr. 2020. Chronic Disaster: Reimagining Noncommunicable Chronic Disease. *American Anthropologist* 22 (3): 639–640.

Garth, Hannah, and Ashanté Reese. 2020. *Black Food Matters*. Minneapolis: University of Minnesota Press.

Gilbert, Jessica L., and Rebekah Williams. 2020. "Pathways to Reparations: Land and Healing Through Food Justice." *Human Geography* 13 (3): 228–241.

Gilmore, Ruth Wilson. 2008. *Golden Gulag.* Berkeley: University of California Press.

Gilmore, Ruth Wilson. 2022. *Abolition Geography*. New York: Verso.

Gottlieb, Robert, and Anupama Joshi. 2010. *Food Justice.* Cambridge, MA: MIT Press.

Guthman, Julie. 2004. *Agrarian Dreams.* Berkeley: University of California Press.

Guthman, Julie. 2008. "If They Only Knew: Color Blindness and Universalism in California Alternative Foods Institutions." *Professional Geographer.* 60 (3): 387–397.

Hartman, Saidiya. 2007. *Lose Your Mother: A Journey Along the Atlantic Slave Route.* New York: Farrar, Straus and Giroux.

Hassberg, Analena Hope. 2020. "Citified Sovereignty: Cultivating Autonomy in South Los Angeles." In *A Recipe for Gentrification,* edited by Alison Hope Alkon, Yuki Kato, and Joshua Sbicca. New York: NYU Press.

Hawthorne, Camilla, and Jovan Scott Lewis. 2023. *The Black Geographic: Praxis, Resistance, Futurity.* Durham, NC: Duke University Press.

Hayes, Kelly, and Mariame Kaba. 2022. *Let this Radicalize You.* Chicago: Haymarket Books.

Hayes-Conroy, Allison, and Jessica Hayes-Conroy. 2010. "Visceral Difference: Variations in Feeling (Slow) Food. *Environment and Planning A* 42 (12): 2956–2971. https://doi.org/10.1068/a4365.

Heynen, Nik. 2009. "Bending the Bars of Empire from Every Ghetto for Survival." *Annals of the Association of American Geographers* 99 (2): 406–422. https://doi.org/10.1080/00045600802683767.

Heynen, Nik, Hilda E. Kurtz, and Amy Trauger. 2012. "Food Justice, Hunger and the City." *Geography Compass* 6 (5): 304–311. https://doi.org/10.1111/j.1749-8198.2012.00486.x.

Heynen, Nik, and Megan Ybarra. 2020. "On Abolition Ecologies and Making Freedom as a Place." *Antipode* 53 (1): 21–35. https://doi.org/10.1111/anti.12666.

Hidalgo, L. (2015). "Augmented Fotonovelas: Creating New Media as Pedagogical and Social Justice Tools." *Qualitative Inquiry* 21 (3): 300–314. https://doi.org/10.1177/1077800414557831.

Hobart, Hiʻilei Julia Kawehipuaakahaopulani, and Tamara Kneese. 2020. "Radical Care Survival Strategies for Uncertain Times." *Social Text* 38 (1): 1–16.

hooks, bell. 2009. *Belonging: A Culture of Place.* New York: Routledge.

Hoover, Elizabeth, and Devon Minhesuah. 2009. *Indigenous Food Sovereignty in the US*. Tulsa: University of Oklahoma Press.

Jernigan, Valerie Blue Bird. 2021. "Ending Food Insecurity in Native Communities Means Restoring Land Rights, Handing Back Control. May 27. The Conversation. https://theconversation.com/ending-food-insecurity-in-native-communities-means-restoring-land-rightshanding-back-control-158858.

Jones, Naya. 2019. "Dying to Eat? Black Food Geographies of Slow Violence and Resilience." *ACME* 18 (5): 1076–1099. https://acme-journal.org/index.php/acme/article/view/1683.

Julier, Alice. 2013. *Eating Together*. Champaign-Urbana: University of Illinois Press.

Keeve, Christian Brooks. 2020. "Fugitive Seeds." *Edge Effects*. https://edgeeffects.net/fugitive-seeds/.

Kelly, Robin D. G. 2022. *Freedom Dreams*. Chicago: Haymarket University Press.

Kimmerer, Robin Wall. 2013. *Braiding Sweetgrass*. Minneapolis, MN: Milkweed Editions.

Kobayashi, A. 2010. "People, Place, and Region." *Annals of the Association of American Geographers* 100 (5): 1095–1106. https://doi.org/10.1080/00045608.2010.523346.

Kobayashi, A. 2014. "The Dialectic of Race and the Discipline of Geography." *Annals of the Association of American Geographers* 104 (6): 1101–1115. https://doi.org/10.1080/00045608.2014.958388.

Kohl, Ellen, and Priscilla McCutcheon. 2015. "Kitchen Table Reflexivity: Negotiating Positionality Through Everyday Talk." *Gender, Place & Culture* 22 (6): 747–763, https://doi.org/10.1080/0966369X.2014.958063.

Kolavalli, Chaaya. 2023. *Well-Intentioned Whiteness*. Athens, GA: University of Georgia Press.

Levkoe, C. Z., C. Hammelman, K. Reynolds, X. Brown, M. J. Chappell, R. Salvador, and B. Wheeler. (2020). "Scholar-Activist Perspectives on Radical Food Geography: Collaborating Through Food Justice and Food Sovereignty Praxis." *Human Geography* 13 (3): 293–304. https://doi.org/10.1177/1942778620962036.

Mares, Teresa, and Devon Peña. 2011. "Environmental and Food Justice." In *Cultivating Food Justice*, edited by Alison Alkon and Julian Agyeman. Cambridge, MA: MIT Press.

Martinez, Aja. 2020. *Counterstory*. St. Louis, MO. National Council of Teachers of English.

McClintock, Nathan. 2011. "From Industrial Garden to Food Desert." In *Cultivating Food Justice*, edited by Alison Alkon and Julian Agyeman. Cambridge, MA: MIT Press.

McClintock, Nathan. 2018. "Cultivating (a) Sustainability Capital." *Annals of the Association of American Geographers* 108 (2): 579–590.

McClintock, Nathan, and M. Simpson. 2017. "Stacking Functions: Identifying Motivational Frames Guiding Urban Agriculture Organizations and Businesses in the United States and Canada." *Agriculture and Human Values* 35, 19–39.

McCutcheon, Priscilla, Latrica E. Best, and Theresa Ann Rajack-Talley. 2023. *Beyond the Kitchen Table: Black Women and Global Food Systems.* Chapel Hill: University of North Carolina Press.

McCutcheon, Priscilla. 2011. "Community Food Security 'For Us, By Us.'" In *Cultivating Food Justice*, edited by Alison Alkon and Julian Agyeman. Cambridge, MA: MIT Press.

McCutcheon, Priscilla. 2019. "Fannie Lou Hamer's Freedom Farms and Black Agrarian Geographies." *Antipode* 51 (1): 207–224. https://doi.org/10.1111/anti.12500.

McKittrick, Katherine. 2006. *Demonic Grounds*. Chapel Hill, NC: Duke University Press.

McKittrick, Katherine. 2015. *Sylvia Wynter: On Being Human as Praxis*. Chapel Hill, NC: Duke University Press.

McKittrick, Katherine. 2021. *Dear Science and Other Stories*. Chapel Hill, NC: Duke University Press.

Meehan, Katie, Mabel Denzin Gergan, Sharlene Mollett, and Laura Pullido. 2023. "Unsettling Race, Nature, and Environment in Geography." *Annals of the American Association of Geographers* 113 (7): 1535–1542.

Meyers, Justin Sean. 2022. *Growing Gardens Building Power*. New Brunswick, NJ: Rutgers University Press.

Minkoff-Zern, Laura-Anne, Nancy Peluso, Sandy Brown and Christy Getz. 2011. "Race and Regulation: Asian Immigrants in California Agriculture." In Alkon, Alison Hope and Julian Agyeman (eds). *Cultivating Food Justice: Race, Class and Sustainability*. Cambridge, MA: MIT Press.

Minkoff-Zern, L. A., and M. A. Carney. 2015. "Latino Im/migrants, 'Dietary Health' and Social Exclusion: A Critical Examination of Nutrition Interventions in California." *Food, Culture & Society* 18 (3): 463–480.

Minkoff-Zern, Laura-Anne, and Teresa M. Mares. 2025. *Will Work for Food: Labor Across the Food Chain*. Berkeley: University of California Press.

Morales, Alfonso. 2011. "Growing Food and Justice." In *Cultivating Food Justice*, edited by Alison Alkon and Julian Agyeman. Cambridge, MA: MIT Press.

Morrison, Dawn. 2011. "Indigenous Food Sovereignty: A Model for Social Learning." In *Food Sovereignty in Canada: Towards a Just and Sustainable Food System*, vol. 6, edited by H. Wittman, Annette Aurélie Desmarais, and Nettie Wiebe. Halifax, Canada: Fernwood.

Morrow, Oona. 2023. *Community Food Initiatives: A Critical Reparative Approach*. New York: Routledge.

Motta, Renatta. 2021. "Social Movements as Agents of Change: Fighting Intersectional Food Inequalities, Building Food as Webs of Life." *Sociological Review*. 69(3): 603-625.

Moulton, A. A., S. Velednitsky, D. M. Harris, C. B. Cook, and B. L. Wheeler, B. L. 2021. "On and Beyond Traumatic Fallout: Unsettling Political Ecology in Practice and Scholarship." *Journal of Political Ecology* 28(1): 677–695.

Muñoz, Lorena, and Megan Ybarra. 2019. "Latinx Geographies." *Society and Space*. https://www.societyandspace.org/forums/latinx-geographies.

Nixon, Rob. 2011. *Slow Violence and the Environmentalism of the Poor*. Cambridge, MA: Harvard University Press.

Norgaard, Kari, Ron Reed, and Carolina Van Horn. 2011. "A Continuing Legacy: Institutional Racism, Hunger and Nutritional Justice on the Klamath." In *Cultivating Food Justice*, edited by Alison Alkon and Julian Agyeman. Cambridge, MA: MIT Press.

Omi, Michael, and Howard Wynant. 2014. *Racial Formations in the United States*. New York: Routledge.

Parasecoli, Fabio. 2022. *Gastronativism*. New York: Columbia University Press.

Parker, Barbara. 2019. *Feminist Food Studies*. Toronto: Canadian Scholars' Press.

Pellow, David. 2016. "Toward a Critical Environmental Justice Studies." *DuBois Review* 13 (2): 221–236. https://doi.org/10.1017/S1742058X1600014X.

Penniman, Leah. 2018. *Farming While Black*. New York: Chelsea Green.

Porter, Christine M., and Alyssa Wechsler. 2018. "Follow the Money: Resource Allocation and Academic Supremacy among Community and University Partners in Food Dignity." *Journal of Agriculture, Food Systems, and Community Development* 8(Suppl. 1): 63–82.

Pulido, L., and J. De Lara. 2018. "Reimagining 'Justice' in Environmental Justice: Radical Ecologies, Decolonial Thought, and the Black Radical Tradition." *Environment and Planning E: Nature and Space* 1 (1–2): 76–98. https://doi.org/10.1177/2514848618770363.

Pulido, Laura. 2017. "Geographies of Race and Ethnicity II: Environmental Racism, Racial Capitalism and State-Sanctioned Violence." *Progress in Human Geography* 41 (4): 524–533.

Ramírez, Margaret Marietta. 2015. "The Elusive Inclusive." *Antipode* 47 (3): 748–759. https://doi.org/10.1111/anti.12131.

Ray, Krishnendu. 2016. *The Ethnic Restauranteur*. New York: Bloomsbury.

Reese, Ashanté. 2019. *Black Food Geographies*. Chapel Hill: University of North Carolina Press.

Reese, Ashanté, and Joshua Sbicca. 2022. "Food and Carcerality." *Food and Foodways* 30 (1–2): 1–15. https://doi.org/10.1080/07409710.2022.2030931.

Reynolds, Kristin, Daniel Block, and Katherine Bradley. 2018. "Food Justice Scholar-Activism and Activist-Scholarship." *ACME* 17 (4): 988–998.

Reynolds, Kristin, and Nevin Cohen. 2016. *Beyond the Kale.* Athens: University of Georgia Press.

Roane, J. T., and Justin Hosbey. 2019. "Mapping Black Ecologies." *Current Research in Digital History* 2. https://doi.org/10.31835/crdh.2019.05.

Robinson, Cedric. 1983. *Black Marxism.* Chapel Hill: University of North Caroline Press.

Sbicca, Joshua. 2018. *Food Justice Now.* Minneapolis: University of Minnesota Press.

Sbicca, Joshua, and Justin Sean Meyers. 2017. "Food Justice Racial Projects: Fighting Racial Neoliberalism from the Bay to the Big Apple." *Environmental Sociology* 3 (1): 30–41, https://doi.org/10.1080/23251042.2016.1227229.

Settee, Priscilla, and Shailesh Shukla. 2020. *Indigenous Foods Systems.* Toronto: Canadian Scholars.

Smith, Bobby II. 2023. *Food Power Politics.* Chapel Hill: University of North Carolina Press.

Sze, Julie. 2020. *Environmental Justice in a Moment of Danger.* Berkeley: University of California Press.

Thompson, Ki'Amber. 2023. "Toward a World Where We Can Breathe." *Annals of the American Association of Geographers* 113 (7): 1699–1710.

Trauger, Amy. 2017. *We Want Land to Live.* Athens: University of Georgia Press.

Tuck, Eve. 2009. "Suspending Damage." *Harvard Educational Review* 79 (3): 409–428.

Van Sant, Levi, Richard Milligan, and Sharlene Mollett. 2020. "Political Ecologies of Race." *Antipode* 53 (1): 629–642.

Vasudevan, Pavithra, Margaret Marietta Ramírez, Yolanda González Mendoza, and Michelle Daigle. (2023). "Storytelling Earth and Body." *Annals of the American Association of Geographers* 113 (7): 1728–1744.

Washington, Karen. 2018. "It's Not a Food Desert, It's Food Apartheid." *Guernica.* https://www.guernicamag.com/karen-washington-its-not-a-food-desert-its-food-apartheid/.

White, Monica Marie. 2019. *Freedom Farmers.* Chapel Hill: University of North Carolina Press.

Whyte, Kyle Pows. 2018. "Food Sovereignty, Justice and Indigenous Peoples: An Essay on Settler Colonialism and Collective Continuance." In *The Oxford Handbook on Food Ethics*, edited by A. Barnhill, T. Doggett, and A. Egan. Oxford, UK: Oxford University Press.

Williams, Justine, and Eric Holt Giménez. 2017. *Land Justice*. Berkeley, CA: Food First Books.

Williams-Forson, Psyche. 2022. *Eating While Black*. Chapel Hill: University of North Carolina Press.

Wittman, Hannah, and Annette Desmarais. 2010. *Food Sovereignty*. Berkeley, California: Food First Books.

Woods, Clyde. 2017. *Development Drowned and Reborn*. Athens: University of Georgia Press.

Zurawski, Erica. 2023. "Materializing the Metaphor: Theorizing the Food Desert as a Socio-Spatial-Legal Instrument in the Production of Space." *Human Geography*. 16 (3): 286–298.

I Dreams

1 A North Star Politics of Emancipation: How Radical and BIPOC Traditions Influence Food Movement Strategies

Antonio Roman-Alcalá

> For enslaved folx fleeing bondage, the North Star marked the way to freedom. . . . It was a constant reminder to freedom-dream. Theory is my North Star: it is a steadfast tool to explain without fluff or gimmicks what I am experiencing. . . . Theory does not solve issues—only action and solidarity can do that—but theory gives you language to fight, knowledge to stand on, and a humbling reality of what intersectional social justice is up against. . . . Theory is a practical guide to understanding injustice historically, the needs of people, and where collective power lives. . . . Without theory, the moveable mountain of injustice and oppression seems too big and immobile, but theory helps us understand that our job is not to move mountains but to outmaneuver them.
>
> —Bettina Love (2019, 121–122)

Food and Farm Movements in Times of Racial Reckoning

Food justice is not new. Nor are the politics of grappling with racism's influence on society. However, recent years could be characterized as a time of "racial reckoning." Donald Trump's 2016 election and an associated right-wing resurgence have reminded a sometimes self-assuredly "postracial" society that racial injustice remains at its foundation. Meanwhile, leftist movements have repeatedly raised long-standing economic and racial critiques, from 2011's Occupy Wall Street through multiple Black Lives Matter mobilizations and resurgent campaigns for labor justice. Within this context, food and farm movements continue to organize for changes to the food system. But in recent decades, and especially since 2016, they are doing so with greater attention to racial inequalities in the food system—and within movements themselves.

During this period of racial reckoning, those who have most been harmed by inequalities have increasingly taken on leadership in food movements. In the US, BIPOC communities have long been present in food movements, especially migrant farmworker organizing, Black agrarianism, and Indigenous struggles for land and food sovereignty (among others). But the critical food studies literature of the 2000s–2010s foregrounded critiques of the predominantly white activists and organizations who were disproportionately funded and lionized in the popular media during that time (Reynolds and Cohen 2013). This scholarship criticized food activists' neoliberal approaches, arguing that their focus on consumption-based strategies, volunteerism, and self-improvement failed to resonate with marginalized communities and limited their potential to make substantial social change (see Guthman 2008, 2011; Alkon and Guthman 2017). A significant portion of this literature, including that on food justice, played up a dichotomy between "oppositional" and "alternative" movement strategies, criticizing the latter as neoliberal and insufficiently "structural" for not directly addressing the state (e.g., Myers and Sbicca 2015; Rosol 2012).

Such critiques deserve credit for pointing out how people of color have too often been excluded from mainstream food and environmental movements and for calling out the dangers of apolitical food reform approaches. But the misinterpretation or misapplication of such critiques may conflate the problems of specific (white, apolitical) movement elements with problems of movement strategy *at large*. Moreover, these critiques are less relevant to today's more BIPOC-inclusive food movements and the ways they have pushed the wider/whiter food movements in more radical directions (Roman-Alcalá 2021). Continuing to rely on these critiques can lead scholars to undervalue the BIPOC sectors whose freedom dreams and everyday work superficially resembles efforts that are subject to the "neoliberal critique" but that are informed by radical analyses of racism, capitalism, and colonialism—and the state, which historically has advanced these systems. In this chapter, I describe this shift toward BIPOC radicalism in food justice movements and what it means for approaches to the state and the market as strategies for emancipatory change. I propose—by synthesizing from movement experience, academic research, and my own analytical perspective—a normative "North Star" politics, which maintains a radical horizon of change beyond capitalism and the state while remaining pragmatic and open to various forms of political action. This North Star is informed by four theoretical linkages—intersectional anticapitalism, Black radicalism, Indigenous resurgence, and anarchism. As Bettina Love describes in the epigraph, theories can help align movement praxis with critical analyses of the systemic injustices at hand and thus guide strategic engagements with the market and the state. And although my thinking about these theories is grounded in countless conversations with and learning from movement participants, the resulting analysis is my own.

What I am calling a North Star politics is a praxis of food system change that is both grounded (in place, in culture, and in history) and expansive (utopian, inclusive, dynamic, and unafraid of "freedom dreams" [Kelley 2003]). It is oriented toward and never undermines transformative outcomes. *And* it engages with flawed systems of capital and state—even while seeking to transcend them. This may sound contradictory, but my research subjects, a broad array of BIPOC food and farming activists including leaders in food banks, agroecological education, Indigenous resurgence, urban farming, migrant dignity, and land access for young farmers, show this isn't impossibly so. The activists I worked with over the course of my seven years of fieldwork, as well as over two decades of food justice activism, are able to navigate difficult tensions—between reforms and reformism, state and market, race and class, institutionalization and autonomy, leadership from above and from below—with grace, openness, and commitment.

Reconsidering State-Facing and Market-Using Strategies

As BIPOC strength and leadership in the food movement has grown, activist-scholars need to (re)consider our approach to the roles of state power and market-based approaches as strategies toward food justice outcomes. The state is often framed dichotomously, with some envisioning it as a source of potential progressive change while others depict it as the central organizing force for oppression. However, most political *actions* fall between these two extremes of embracing and rejecting the state, and indeed I have found strategic nuance in how food justice movement actors address this question. Similarly nuanced are the ways that BIPOC food activists think about, approach, and address markets.

In the context of prevalent right-wing antistate discourses and neoliberalism, some scholar-critics have judged antistate and market-utilizing approaches as apolitical, unstrategic, or even politically regressive. Yet the

BIPOC radicalism I observe in my fieldwork is rooted in generations of victimization at the hands of the state and is decidedly political, strategic, and progressive. Moreover, the activists I worked with recognize that the post-WWII "golden age" in which working class wages improved, government programs supported social welfare, and the state brokered a relative peace between organized labor and capital was short-lived, racially uneven, and brought about only because it was flanked by domestic socialist agitation and the threat of global communist revolution. Only renewing a widespread desire for alternatives to capitalism can recreate the US state's return to its Keynesian heyday.

In this context, tactics implicated in critiques of neoliberalism can support a *politicized* vision of a future without capitalism and with a deeper democracy beyond the corrupt nation-state. In my research, I found that the growing voice and influence of BIPOC sectors in food movements has led to more radical theoretical frames, including critiques of capitalism, that are nonetheless combined with market-leveraging power-building strategies. The racial, colonial, capitalist state remains a target of politics for pragmatic reasons but is approached skeptically and does not define the emancipatory horizon for future political arrangements.

Intersectional Anticapitalism

Many of the activists I worked with agree with prominent Black farmer-educator Leah Penniman, who affirms, "We need to question capitalism itself" (Retta 2021). Generally, food and farm movements are more liberal-reformist than revolutionary, and critiques of "capitalism" can sometimes dress up what is in actuality a critique of corporate power. But some food justice actors do implicate capitalism as a systemic driver of injustice and see the state as structurally beholden to the owning class, perspectives that imply the need for radical change.

Food movement critiques of capitalism often reflect what Cedric Robinson refers to as "racial capitalism," meaning that racial and economic exploitations are not independent forces but inseparably interlinked (Robinson 2000 [1983]; Moten 2018). The activists I've worked with name capitalism, in its intersections with racism and colonialism, as creating the conditions they are fighting against. For example, the 2018 National Assembly of the multiethnic US Food Sovereignty Alliance describes its 100+ organizational members as "fighting for their livelihoods, freedom, and dignity through food sovereignty, and [for] an end to patriarchal, capitalist, and colonial ways of thinking and acting." At the local level, Black Star Farmers, a gardening and educational initiative that grew out of antiracist protests in Seattle in 2020, is "committed to facilitating difficult conversations about race, anti-colonialism, anti-capitalism, and the abolition of all oppressive systems especially as it relates to BIPOC communities" (Black Star Farmers 2022). Another recent initiative, Minnow, "secures land tenure for California's farmers of color and indigenous communities to support the heritage and foodways of those most affected by state-sponsored dispossession. Such conditions help repair the harms of settler colonialism, which persist to this very day under capitalism and white supremacy, [and] supports land relations that transcend the capitalist logic of land as commodity" (Minnow n.d.).

Even traditionally whiter, larger, and more established groups like the National Young Farmers Coalition (NYFC) have increasingly incorporated BIPOC membership and influence and begun to describe their goals in more radical terms. Their 2022 national member survey affirmed that "challenging capitalism is one of the National Young Farmers Coalition's Guiding Principles" (NYFC 2022). Elsewhere, NYFC describes the intersectional nature of their anticapitalism, the "historic and ongoing violence against Black, Indigenous, Latine, and other communities of color" within American agriculture, and the need to evolve or abolish these systems to achieve collective liberation (NYFC n.d.).

Many of these projects utilize business structures and selling food at markets while seeking to build alternatives to capitalist logics. The Federation of Southern Cooperatives, for example, is a decades-old southeastern Black farmer coalition that supports local cooperatives and is a member of the avowedly anticapitalist global peasant movement La Vía Campesina. Namu Farms and Feral Heart Farm are Asian American–owned businesses that name capitalism as a root of cultural dispossession and food injustice. Feral Heart's owner told me he dreams of "subverting the dominant capitalist farming business model to truly transform our food pathways." Planting Justice is a nonprofit nursery and food-growing business that links food justice to efforts to abolish the prison industrial complex. They employ formerly incarcerated people while building capacity to fight local struggles, such as a campaign against government funding of jails rather than jail-to-work programs (Zandi 2009). All of these organizations buy and sell local organic food as a means to more collective and radical aims.

How might politicized food/farm businesses lead to overcoming capitalism? What are the limits of making transformative change through business? The projects and individuals with anticapitalist ideals who I've encountered for decades offer no easy answers. And yet, anticapitalist desires remain a North Star, guiding the visions and work of these BIPOC activists.

Black Radicalism

The Black Radical Tradition (BRT) encompasses a collection of intellectual, cultural, and activist work that resists structures and ontologies rooted in slavery, imperialism, and capitalism (Robinson 2000[1983]). Within this tradition, Afropessimism argues that Blackness is defined in opposition to humanness, preventing the political subjecthood of Blacks even within the liberal-assimilationist state (Hartman 2007; Sharpe 2016). Afropessimism is resolutely antistate, seeing the state as a central architecture for maintaining Black subjugation (Samudzi and Anderson 2018). In this context, the BRT and Afropessimism envision emancipatory political formations based on the figure of the maroon—formerly enslaved individuals who escaped and created their own hidden, self-reliant settlements, often in conjunction with Indigenous peoples, on the outskirts of colonization's frontiers (Wright 2020; Bledsoe 2017; Simpson 2017). Although physical marronage may seem a remote possibility, today's Black farmers and agrarian communities build upon this history by emphasizing collectivity, self-determination, and autonomy (White 2018, 6; Garth and Reese 2020; chapter 4, this volume).

Black Earth Farms, for example, insists on the need for "autonomous and sovereign communities that steward and produce their own ethically harvested food" (n.d., paragraph 3). The group is clear about its targets: "Anti-Black racial capitalism, Anti-Indigenous Imperialist Empires, and white supremacist hetero-patriarchal structures, practices, and ongoing projects." They call for these structures, along with "pseudo socialism and fascist neoliberalism" to "be abolished and replaced with autonomous and sovereign communities centered around indigenous self-determination and Black liberation" (2021, paragraph 3). We can see similar prefigurative efforts to build collective autonomy from racial capitalism in groups like Minnow, which seeks to restore BIPOC access to land without indebtedness.

Amid the industrial agriculture of California's Central Coast is Tierras Milperas, a farmworker-led organization that operates several community gardens. They describe their desire for self-determination as rooted in histories of oppression, and their vision, as stated on their GoFundMe page, includes independence from the nonprofit sector as well as the state: "It is a collective emergence from our campesino histories, culture and the reclamation of our autonomy to decide for our future by not forgetting our past. Three years ago our community . . . became independent from the practices of control of a local non-profit. Since then we have steered the direction and care of 7 community gardens . . . as a community." Here we see how autonomy, a rejection of the "nonprofit industrial complex," and a focus on collective existence underpin a community-level, alternative economy project.

The Black Radical Tradition condemns the racial capitalist, settler colonial state's inability to recognize the humanity of Black communities and extend to them political subjecthood. Theorists and practitioners find emancipatory models in the histories of maroon communities creating fugitive and even joyful existences on the geographic and social margins. Radical BIPOC food activists draw from this tradition when they emphasize collective autonomy from the state and the nonprofit sectors as their guiding North Stars. Rather than craft neoliberal subjectivities, they seek to enact prefigurative collective and utopian liberatory imaginaries.

Indigenous Resurgence

Indigenous theorizing and struggles in so-called North America also illustrate distrust for the state and a desire to build communal autonomy and economic self-reliance. Like the BRT, Indigenous theory emerges from particular experience and is therefore not unproblematically applied to non-Indigenous groups. Still, these ideas have shaped US history and culture and can inform transformation-seeking politics.

One common thread among radical Indigenous theorists is an emphasis on *relationality* and an ethic of *obligation to all relatives* (Coulthard and Simpson 2016; Romero-Briones et al. 2020; chapter 3, this volume). This relational ethic is ontologically opposed to the extractive, distancing, instrumental profit-seeking ethics of capitalism. Indigenous theory shares Black agrarians' practical consideration of land as a means to collective autonomy and sovereignty. For example, the Cultural Conservancy describes its Heron Shadow farm, which distributes food freely to various tribes, as an "Indigenous Biocultural Heritage Oasis . . . that will renew and revitalize Native landscapes, traditional foods, and Indigenous lifeways [in order to] support tribal sovereignty, reciprocal exchange and trade, and Native food systems" (Cultural Conservancy, n.d., 3–5). Moreover, Indigenous relationality moves land and food politics from mere material to cosmological. It points to a nonpossessive politics of belonging in which the active practice of land relations is central.

Indeed, the #LandBack movement states clearly that decolonization demands changes to land ownership and use in addition to epistemological and cultural recognition (Pieratos et al. 2020; Kauanui 2016; chapter 5, this volume). I encountered diverse engagements with property among Indigenous-led organizations. The urban Sogorea Te' Land Trust worked with one city government to defend a sacred site from development and desecration while fighting *against* another city government for the same purpose. And Indigenous projects differ in their gathering practices for ancestral foods; some work with private landowners, while others harvest plants illicitly. Still others work with land trusts to decommodify land within colonial-capitalist legal-economic structures. While they use different approaches, all of these groups are following their North Stars toward decommodified Indigenous land for Native food practices.

Indigenous theory speaks to a centuries-long struggle not simply *against* colonization but for a *continuity* of cultural tradition and place-based belonging—an assertion of Indigenous pride and strength rather than a mere reaction to an imposition (Corntassel 2021). Land and food sovereignty are essential aspects of this continuance (Whyte 2018). Indigenous food projects and organizations engage with the market and the state as they deem it necessary to realize this vision.

Anarchism

Anarchism privileges the decentralization of political power and endogenous community development—and thus rejects the state. Anarchists prefer to address issues and build power through prefigurative strategies. The resurgence of anarchist approaches can be seen in the explosion of "mutual aid" projects after COVID-19 (Roman-Alcalá 2020; Spade 2020). Alongside these projects' voluntaristic, decentralized structure was a widely adopted rejection of charity and assertion that grassroots care and solidarity are the ways we "build a new world in the shell of the old" (as goes the old anarcho-syndicalist saying).

These anarchist principles resonate with the lineages outlined above. They overlap with the BRT and Afropessimism in a historically informed assessment of the state as oppressor and in focusing on escape and attack strategies (Anderson 2021; Bey 2020). Reflecting anarchism's prefigurative ideals, Raised Roots Farm's Jamil Burns described his work to feed frontline protestors in the wake of George Floyd's murder: "I think a lot of people have woken up not only to racial injustices but to the idea that in order for a community to be truly resilient, we have to start relying on each other." Anarchism takes criticism of the alienating, procapitalist, and developmentalist state to its conclusion by proposing that both the means and ends for emancipation are outside and beyond nation-states (Corntassel 2012; Lewis 2017).

Rejection of state politics and the pursuit of other means is apparent in autonomous garden spaces across so-called North America, from the 1969 creation (and ongoing defense) of Berkeley's People's Park, to Black autonomous gardens in Detroit, to queer/BIPOC land projects in gentrifying Atlanta like Mariposas Rebeldes and its opposition to "Cop City," a proposed police training facility. Such gardens involve a wide array of identities and backgrounds and are not always explicitly anarchist. But some are, such as the Food Autonomy festival put on as part of the Stop Cop City movement (ACPC 2023). Manuel "Tortugita" Terran, murdered by Atlanta police while occupying the Cop City site, was an anarchist.

Anarchists critique leftist tendencies that seek to capture state power, holding that those pursuing radical change through the state will be shaped into enterprising reformists, compromisers, and willing elites. Famous early anarchist theoretician Peter Kropotkin long ago called this "the conquest of socialists by power," and modern anarchist theorists point out myriad examples in practice (Baker 2019). Instead, anarchists decry reformism as a counterinsurgency that can validate a system *premised on inequality* while reducing investment among movements in prefigurative and autonomy-building approaches that are potentially—though not certainly—emancipatory (Rodriguez 2020).

Pragmatic food justice activists struggle to use the state to their own ends without validating the system itself. A member of the educational cooperative

Agroecology Commons described their interest in state funds as "shaking the money tree," an approach common to many BIPOC food activists in California.[1] They do not completely reject the state but approach an interest in state funds and support with skepticism and caution—and sometimes, as reparations. Anarchism thus offers an ethic of political action that may not always *deny* state-facing action, even while it certainly demotes it. The Mexican/Indigenous-descended director of a food bank in California's Central Valley, who described herself as an anarchist, exemplifies this tension. "Revolution happens in many, many different ways and with many, many different tactics, and so I am not here to bash any other type of movement-making that is going on," she said. "Whatever their work is, I'm here to support it. As long as we're moving towards justice and equity I'm down."[2] This activist, and others I met during my research, did not universally deny statist strategies but put the bulk of their efforts into other means of building food justice and BIPOC power.

Following the North Star

These examples of radical yet pragmatic BIPOC activism push activist-scholars to question why movements move as they do and why states act as they do, which together may provide theoretically informed strategic guidance to food justice movements, including both its mainstream and radical flanks.

What might such theories describe? For one, they would recognize the representative, liberal, politically alienating state as a foundation, *alongside racial capitalism and colonialism*, for social injustice. This description supports strategic skepticism of the state, which will not be our savior but still cannot (often/always) be ignored. Food justice analysis that is more forthright and sober about the problematic role of the state may temper the tendency of academics and nonprofit professionals to prioritize the state as the key avenue toward transformation. It could thus open up analysis and work toward other levers of change that either sidestep the state or bolster the likelihood that state-oriented strategies would prove useful by strengthening movements outside it (as we see in today's movement against Cop City). It is one thing to argue that the state is a "necessary evil," but it is another thing—and unfortunately a much more common thing—to unproblematically reinforce the state's hegemony as a locus of politics. In doing the latter, researchers risk undermining the radical freedom dreams of BIPOC movements for emancipation. Instead, centering the importance of ambitious, "unrealistic" dreams driven by hope, love, and dignified rage helps to support radicalism in the movement and radical change in society.

Recognizing BIPOC radicalism can also expand patterns of solidarity within movements, countering the long-standing tradition of conflating threats to the established (racial) order with a demonized communism (Burden-Stelly 2017). Even those who do not hold racialized or radical identities, backgrounds, or proclivities can support their allies in struggle by accepting a diversity of tactics, such as recognizing the right to self-defense and the usefulness of destructive tactics such as disruption, blockades, sabotage, and illegal appropriation. Radicalism also reminds us of the benefits of alternative institutions and political formations outside the mainstream. Acceptance of radicalism may lead to even better strategic synergies between sectors and initiatives.

State skepticism, as we find among *some* (but not all) environmental justice, Indigenous, Black, and migrant organizing, does not preclude taking opportunities that arise regarding the state. Indeed, as long-standing targets of state violence, BIPOC communities know full well the imperative to deal with the state. But skepticism can guide actors to take available opportunities in a way that subverts cooptation, demobilization, and other negative dynamics of movement institutionalization (e.g., Liévanos 2012). Together, state skepticism and anticapitalism enable approaches that strengthen movement actors' collective agency while mitigating the corrupting and compromising influences of the state and the market. This is a delicate balance, of course, since the benefits of state support and the need to survive within capitalist labor markets are undeniable. But moving toward North Stars of emancipation while maintaining that balance remains a possibility.

Notes

1. Interview, February 2021.
2. Interview, September 2020.

References

ACPC (Atlanta Community Press Collective). 2023. "'We Are Not in the Least Afraid of Ruins': Food Autonomy in the Weelaunee Forest." https://atlpresscollective.com/2023/03/19/we-are-not-in-the-least-afraid-of-ruins-food-autonomy-in-the-weelaunee-forest/.

Alkon, A. H., and J. Guthman, eds. 2017. *The New Food Activism: Opposition, Cooperation, and Collective Action.* Oakland: University of California Press.

Anderson, W. C. 2021. *The Nation on No Map: Black Anarchism and Abolition.* Oakland, CA: AK Press.

Baker, Z. 2019. "Means and Ends: The Anarchist Critique of Seizing State Power." Black Rose/Rosa Negra. https://blackrosefed.org/anarchopac-critique-of-seizing-state-power.

Bey, M. 2020. *Anarcho-Blackness: Notes Toward a Black Anarchism.* Chico, CA: AK Press.

Black Earth Farms. n.d. "Organizational Principles and Manifesto." Accessed May 26, 2021. https://web.archive.org/web/20210604041805/https://www.blackearthfarms.com/.

Black Star Farmers. 2022. "SLEE 2022 Keynote by Black Star Farmers: Reclaiming Food Sovereignty and Reconnecting to Indigenous Ways of Life." https://cagj.org/slee-2022/keynote-black-star-farmers/.

Bledsoe, A. 2017. "Marronage as a Past and Present Geography in the Americas." *Southeastern Geographer* 57 (1): 30–50.

Burden-Stelly, C. 2017. "Cold War Culturalism and African Diaspora Theory: Some Theoretical Sketches." *Souls* 19 (2): 213–237. https//doi.org/10.1080/10999949.2016.1239169.

Corntassel, J. 2012. "Re-Envisioning Resurgence: Indigenous Pathways to Decolonization and Sustainable Self-Determination." *Decolonization, Indigeneity, Education and Society* 1 (1): 86–101.

Corntassel, J. 2021. "Life Beyond the State: Regenerating Indigenous International Relations and Everyday Challenges to Settler Colonialism." *Anarchist Developments in Cultural Studies* 21 (1): 71–97.

Coulthard, G., and L. Betasamosake Simpson. 2016. "Grounded Normativity/Place-Based Solidarity." *American Quarterly* 68 (2): 249–255.

Cultural Conservancy. n.d. Heron Shadow. Accessed March 31, 2025. https://www.nativeland.org/heron-shadow.

Guthman, J. 2008. "Neoliberalism and the Making of Food Politics in California." *Geoforum* 39 (3): 1171–1183. http://doi.org/10.1016/j.geoforum.2006.09.002.

Guthman, J. 2011. "'If They Only Knew': The Unbearable Whiteness of Alternative Food." In *Cultivating Food Justice: Race, Class, and Sustainability*, edited by A. H. Alkon and J. Agyeman, 263–282. Cambridge, MA: MIT Press.

Hartman, S. 2007. *Lose Your Mother: A Journey Along the Atlantic Slave Route.* New York: Farrar, Straus and Giroux.

Kauanui, J. K. 2016. "'A Structure, Not an Event': Settler Colonialism and Enduring Indigeneity." *Lateral* 5 (1). https://doi.org/10.25158/l5.1.7.

Kelley, R. 2003. *Freedom Dreams: The Black Radical Imagination.* Boston: Beacon Press.

Lewis, A. G. 2017. "Imagining Autonomy on Stolen Land: Settler Colonialism, Anarchism and the Possibilities of Decolonization?" *Settler Colonial Studies* 7 (4): 474–495. https://doi.org/10.1080/2201473X.2016.1241211.

Liévanos, R. 2012. "Certainty, Fairness, and Balance: State Resonance and Environmental Justice Policy Implementation." *Sociological Forum* 27 (2): 481–503. https://doi.org/10.1111/j.1573-7861.2012.01327.x.

Love, B. 2019. *We Want to Do More Than Survive: Abolitionist Teaching and the Pursuit of Educational Freedom.* New York: Beacon Press.

Minnow. n.d. "About." Accessed March 31, 2025. https://www.weareminnow.org/about-1.

Moten, F. 2018. "A Conversation with Fred Moten 12/02/18." YouTube. https://www.youtube.com/watch?v=I6b5N_u7Ebs.

Myers, J. S., and J. Sbicca. 2015. "Bridging Good Food and Good Jobs: From Secession to Confrontation within Alternative Food Movement Politics." *Geoforum* 61: 17–26. http://doi.org/10.1016/j.geoforum.2015.02.003.

NYFC. n.d. "Guiding Principles." Accessed March 31, 2025. https://www.youngfarmers.org/guiding-principles/.

NYFC. 2022. "Building a Future with Farmers 2022: Results and Recommendations from the National Young Farmer Survey." https://www.youngfarmers.org/wp-content/uploads/2022/08/NationalSurveyReport2022.pdf.

Pieratos, N. A., S. S. Manning, and N. Tilsen. 2020. "Land Back: A Meta Narrative to Help Indigenous People Show up as Movement Leaders." *Leadership* 17 (1): 47–61. https://doi.org/10.1177/1742715020976204.

Retta, M. 2021. "How to Provide for Your Neighbors and Yourself in a Racist Food System." *Vice.* https://www.vice.com/en/article/epn49n/how-to-provide-for-your-community-and-yourself-in-a-racist-food-system.

Reynolds, K., and N. Cohen. 2016. *Beyond the Kale: Urban Agriculture and Social Justice Activism in New York City.* Athens: University of Georgia Press.

Robinson, C. 2000 [1983]. *Black Marxism: The Making of the Black Radical Tradition.* London: Zed Press.

Rodriguez, D. 2020. *White Reconstruction: Domestic Warfare and the Logics of Genocide.* Minneapolis: University of Minnesota Press.

Roman-Alcalá, A. 2020. "Op-Ed: We can Build a Better Food System Through Mutual Aid." Civil Eats. https://civileats.com/2020/06/26/op-ed-we-can-build-a-better-food-system-through-mutual-aid/.

Roman-Alcalá, A. 2021. *Histories of Othering, Practices of Solidarity, and Prospects for Emancipatory Convergence Among California's Food and Farming Movements in Times of Resurgent Rightwing Power*. PhD thesis zine, International Institute of Social Studies, Erasmus University Rotterdam. https://www.iss.nl/en/media/2021-12-antonio-roman-alcala-diss-zine.

Romero-Briones, A., E. Salmon, H. Renick, and T. Costa. 2020. "Recognition and Support of Indigenous California Land Stewards, Practitioners of Kincentric Ecology." First Nations Development Institute, https://www.firstnations.org/wp-content/uploads/2020/08/Indigenous-California-Land-Stewards-Practitioners-of-Kincentric-Ecology-Report-2020.pdf.

Rosol, M. 2012. "Community Volunteering as Neoliberal Strategy? Green Space Production in Berlin." *Antipode* 44 (1): 239–257. http://doi.org/10.1111/j.1467-8330.2011.00861.x.

Samudzi, Z., and W. C. Anderson. 2018. *As Black as Resistance: Finding the Conditions for Liberation*. Chico, CA: AK Press.

Sharpe, C. 2016. *In the Wake: On Blackness and Being*. Durham, NC: Duke University Press.

Simpson, L. B. 2017. *As We Have Always Done: Indigenous Freedom through Radical Resistance*. Minneapolis: University of Minnesota Press.

Spade, D. 2020. *Mutual Aid: Building Solidarity during this Crisis (and the Next)*. London: Verso.

White, M. 2018. *Freedom Farmers: Agricultural Resistance and the Black Freedom Movement*. Chapel Hill: University of North Carolina Press.

Whyte, K. P. 2018. "Food Sovereignty, Justice, and Indigenous Peoples: An Essay on Settler Colonialism and Collective Continuance." In *The Oxford Handbook of Food Ethics*, edited by A. Barnhill, M. Budolfson, and T. Doggett, Oxford University Press, pp. 345–366.

Wright, W. J. 2020. "The Morphology of Marronage." *Annals of the American Association of Geographers* 110 (4): 1134–1149. https://doi.org/10.1080/24694452.2019.1664890.

Zandi, H. 2009. "A Common Ground Found through Food Justice: Growing Coalitions between Environmental, Human Rights, Anti-Imperialism, and Labor Movements." https://web.archive.org/web/20230531002051/https://plantingjustice.org/planting-justice/resources-food-justice-research-a-common-ground-found-through-food-justice/.

2 Queering Food Justice: Confronting White Heteropatriarchy in US Agriculture

Michaela Hoffelmeyer and Isaac Sohn Leslie

LGBTQ+ farmers have always existed, dreaming of futures unbounded by heteropatriarchy. In the past decade, farmer-activists and researchers have documented heterosexism and cissexism in farming, LGBTQ+ food insecurity, and queer resistance strategies. Yet despite being leaders of liberation through food systems, the food justice movement has rarely meaningfully engaged with queer or trans thought or politics. Why might this be the case? In this chapter, we look at food justice with a queer theoretical lens to explore how the movement relates to heterosexism, cissexism, and queer resistance in US agriculture.

What does it mean "to queer" food justice? "Queering" means questioning what gender, sexual, and intersecting relations have to do with various aspects of life. In this case, we examine food justice with a queer lens to unpack how intersections of heteropatriarchy and racism may have hindered the movement from achieving its own goals. Queering further encourages us to highlight queer resistance and joy, so we spend the second half of this chapter learning from queer ways of reorganizing farm life in the pursuit of justice.

Food justice scholars and activists have long pointed out that food production and consumption go hand-in-hand in terms of justice, so while this chapter is focused on agriculture, it is important to mention that LGBTQ+ people do not have equitable access to food. Whereas the overall food insufficiency rate (sometimes or often not enough to eat in the past seven days) in the US is 9.5 percent, the number jumps to 15.6 percent for lesbian, gay, and bisexual people and to 22.3 percent for transgender people (Leslie, Carson, and Bruce 2023). Breaking down these numbers by race reveals greater

This chapter has equal authorship.

disparities, with, for instance, 14.7 percent of white transgender people experiencing food insufficiency, compared to 30.7 percent of Hispanic transgender people and 33.3 percent of Black transgender people in New England (Leslie, Carson, and Bruce 2023). These disparities exist because of systemic oppression that affects income (e.g., job and housing discrimination), but they persist even after controlling for income, suggesting that heteropatriarchy and racism operate both through income and outside of income to create these patterns of queer food injustice (Leslie, Carson, and Bruce 2023).

Like food insecurity data, quantitative studies of LGBTQ+ farmers are stymied by the politicization of queerness, contributing to resistance to including sufficient gender identity and sexual orientation questions on large and government-sponsored food and agricultural surveys. Without these data, one study documents 23,701 US farmers in same-sex marriages (Dentzman et al. 2021, 232), a highly conservative estimate due to not including, for example, farms run by single or more than two people, farmers not represented by gender-binary survey questions, and farm workers. Yet queer rural sociologists have begun documenting quantitative evidence of heteropatriarchal discrimination in agriculture and are calling on the US Department of Agriculture and others to count LGBTQ+ farmers as an accountability mechanism for equity (Hoffelmeyer, Wypler, and Leslie 2023).

In this chapter, we offer examples of how a queer lens on core food justice topics contributes to the movement's goals. This work is informed by our interviews, surveys, literature reviews, feedback from presentations, and ongoing community engagement with LGBTQ+ farmers and activists. It is shaped by our positionalities; Michaela is a white, cisgender, queer woman, and an Assistant Professor of Public Engagement in Agriculture at the University of Wisconsin-Madison. Ike is a white, Ashkenazi, trans, and queer owner of Magnetic Fields Farm, LLC (see figure 2.1), and an Extension Assistant Professor of Community Development at the University of Vermont.

We begin by exploring the relationships between heteropatriarchy and white supremacy in US farming through (1) the settler colonial state's use of the heteropatriarchal "family farm" model to hold and transfer private land, (2) heteronormative pathways for farm viability, and (3) white heteronuclear respectability politics in food justice movement organizing. We then offer three examples of queer strategies for realizing food justice movement goals: (1) the "chosen family farm" as an alternative to the (blood) "family farm," (2) LGBTQ+ farmer organizing centered on bodily sovereignty, queer ecology, and queer joy, and (3) Indigenous and anti-colonial queer approaches that question privatized land ownership and develop alternative land relations for liberation. Through these examples, we argue that the intersecting ideologies and material manifestations of white supremacy and heteropatriarchy have contributed to the food justice movement overlooking queer marginalization and resistance strategies across the food system. By examining how LGBTQ+ communities in agriculture disrupt common anti-intersectional notions of queerness as whiteness and food justice as racial justice, we offer queer lenses and strategies for realizing food justice movement goals in agriculture.

Heteropatriarchy Meets White Supremacy in US Farming

In this section, we scrutinize intersections of heteropatriarchy and racism in (1) the family farm: the heteropatriarchal roots of the organizing structure of US farming, (2) farm viability: how heteropatriarchy affects economic outcomes on farms today, and (3) respectability politics: historical heteropatriarchy in food justice activism.

The Family Farm: The Heteropatriarchal Roots of the Organizing Structure of US Farming

In the US, the family farm represents 98 percent of all farms (Whitt, Miller, and Olver 2022). The definition of the family farm is based on farm operators' marriage status and biological relations (Whitt, Miller, and Olver 2022). Defined by operator relations rather than agricultural practices, the category of "family farms" offers little to no indication of social or ecological outcomes that may occur as a result of this model (Hoffelmeyer 2021; Leslie, Wypler, and Bell 2019). Rather, the presumption that family farms offer positive social outcomes enables continued assaults on equity and justice in the food system. For example, Babb et al. (2023) analyze the case of a farmers market where certain community members and vendors leveraged the family farm ideal to allow the continued participation of a racist market vendor. Invoking the "family farm" enabled key actors to dismiss concerns about the violence and exclusion this seller brought to the market. Leveraging the family farm to excuse vendor actions prohibited a more equitable and just response that would have moved toward the progressive ideals supposedly upheld by local and alternative agriculture movements. Despite critiques of the family farm ideology, celebrating the "family farm" is a rare political consensus on the political right and left. This valorization exists in conventional industrial agriculture as well as in organic and other forms of alternative agriculture.

Feminist scholars have demonstrated how the family farm marginalizes women and devalues their labor—both productive and reproductive—in agriculture (Whatmore 1991). Rather than being viewed as "farmers," women, even when active in farm labor and business, have traditionally been considered "farmwives" (Sachs 1983). This demarcation creates barriers for women to access training and resources necessary for farm viability (Sachs et al. 2016). Men make up 64 percent of all producers in the US (USDA NASS 2019). However, when considering the "primary" producer—the person with greater managerial and decision-making power[1]—men make up 76 percent of all producers (USDA NASS 2019). Historically, the family farm was not understood as an equal partnership. Rather, this model reinforces gender norms that further embed gender inequity in agriculture. A queer lens draws on this feminist critique to document how the family farm relies on inequities not only in sex but in gender identity and sexuality as well.

A queer lens reveals how the family farm is a way of organizing land access and farm businesses around a business partner who is also a sexual or intimate partner (Leslie 2017). When we look at the history of the family farm in the US from a queer perspective, we see that farms were not always organized around such heteronuclear family units (Rosenberg 2016). Rather, the family farm model that we have today was actively pushed by the US settler colonial state—through agricultural and social programs like 4-H—in large part to promote rural white repopulation, which maintains control of Indigenous lands (Rosenberg 2016). The family farm is designed to hold land in family lineages, contributing to 98 percent of US farmland being owned by white people today (Horst and Marion 2019).

A queer perspective on the family farm unveils the power relations in that organizational unit and critiques farm structures predicated on privatized land ownership. It reveals how the ownership structure of the family farm depends on disparities between farmworkers and farm owners (Gray 2014). Because the family farm is out of reach or out of interest for many LGBTQ+ farmers (Leslie 2019), a queer perspective calls on us to question the family farm as an ideal way of modeling farming and family.

A queer perspective further encourages us to question a core underpinning of heteronormativity and the family farm: monogamy. Given high divorce rates, monogamy puts farms at risk. From our perspective, a queer lens begs us to ask, *Is monogamy the monoculture of the farm household?* By questioning the very fundamentals of the family farm—including monogamy and the colonial sex-gender binary—a queer lens reveals the family farm as a continued tool of land control and farm organization. It does not suggest that we should simply do away with the family farm but, rather, that an uncritical celebration of it is misguided, and we must create options for land and farm organization beyond the colonial, heteropatriarchal blueprint of the family farm, as we discuss below.

Farm Viability: How Heteropatriarchy Affects Economic Outcomes on Farms Today

Qualitative research that brings a queer perspective to bear on farm viability reveals how most farmers utilize the family farm model to access what Leslie (2019) calls "key resources for farm viability:" land, labor, credit, knowledge, and health care. Farmers often depend on family to access land through wealth, land inheritance, and/or or a partner's off-farm income (Leslie et al. 2019). Farmers regularly rely on intimate partners for on-farm labor, off-farm income, health insurance, and/or credit-building. Yet LGBTQ+ people are less likely to be able to achieve—or be interested in—the family farm model, often leaving these common pathways to farm viability out of reach (Leslie et al. 2019; Hoffelmeyer 2020). Additionally, LGBTQ+ people are more likely to have experienced prior job discrimination and to live in poverty, and so have lower credit scores, reducing their ability to get loans (Badgett et al. 2019; McFadden 2020; Watson et al. 2021). Furthermore, spaces where farmers learn to farm—such as apprenticeships and farmer networkers—are often isolating for LGBTQ+ people because of the lack of LGBTQ+ community there (Wypler 2019). In short, these key resources for farm viability are typically accessed through heteronormative and cisnormative pathways, which reduce LGBTQ+ farmers' ability to achieve farm viability at numerous points in the farm-building process.

Emerging quantitative research finds that LGBTQ+ farmers' most significant challenges are common to all farmers but exacerbated by heteropatriarchal interactions and institutions. These challenges include profitability, land access, affordable and/or available housing, health insurance, and labor (Hoffelmeyer, Wypler, and Leslie 2023). With the addition of housing—another common area of heteropatriarchal discrimination (Movement Advancement Project 2019)—these areas align with the key resources for farm viability described above as typically achieved through heteronormative pathways, and so may be understood as areas of systemic discrimination. Yet when Hoffelmeyer, Wypler, and Leslie (2023) asked LGBTQ+ farmers what challenges they encountered related to their queerness, another top set of findings emerged: anticipated discrimination, social isolation, training opportunities and/or lack of skill, and family dynamics. These are areas of interpersonal discrimination that negatively affect farm viability. Hoffelmeyer, Wypler, and Leslie (2023) contend that interpersonal discrimination is more obviously due to heteropatriarchy than areas of systemic discrimination, and so when asked about their experiences of discrimination, LGBTQ+ farmers are more likely to think of interpersonal rather than systemic examples (such as in accessing key resources for farm viability).

Respectability Politics: Historical Heteropatriarchy in Food Justice Activism

The Black feminist concept of respectability politics reveals how this approach has influenced the food justice movement. Respectability politics is the process of groups or individuals attempting to enact and perform norms associated with a privileged group to make socioeconomic gains or avoid violence. Respectability politics was conceptualized in response to white heteropatriarchal dominance based on efforts to push back against racist stereotypes to establish one's "respectability" (read as proximity to whiteness) (Higginbotham 1993; Cooper 2017). Higginbotham (1993) used the term "politics of respectability" to describe how African Americans were encouraged to present themselves with cleanliness, politeness, high morality, and sexual purity. This approach to gaining advances in the context of racism and anti-Blackness meant that individuals and communities could "achieve" respectability by modeling themselves based on an ideal type of white, middle-class heteronormativity. Conservative sexual politics are baked into respectability politics (Cooper 2017). The norms that construct queerness and transness as gender and sexual deviance mark queerness and transness as things that should not be discussed, acknowledged, or embodied. To be considered "respectable," one must claim white heterosexual and cisgender norms.

In applying a respectability politics lens to the prominent US farm labor union the United Farm Workers (UFW), Minian (2013) shows how in agriculture, respectability politics shapes movements aimed at achieving social justice. Minian (2013) analyzes Cesar Chavez's speeches and the UFW's magazine to show how the UFW embraced heterosexism and heteronormativity to appeal to both white audiences and the predominantly Latino community the organization served. First, Chavez and the UFW spoke out against birth control. Reframing the debate to focus on "the family," the UFW gained support from others who may have subscribed to the negative stereotypes around Mexican families while limiting women's reproductive rights. Second, in an effort to respond to racist stereotypes depicting Mexicans as an overly sexual group, the UFW promoted farmworkers as restrained and respectable—stressing abstinence. Finally, UFW publications portrayed growers and another union as queer, thereby promoting heterosexism and embedding whiteness with queerness. Overall, Minian's (2013) research shows how the UFW erased and silenced the needs of union members by aiming to align with white heteropatriarchal dominance, arguably undercutting their own efforts for economic and social equity.

We question if and how respectability politics may function in today's food justice movement. In attempting to understand why and how the needs of LGBTQ+ communities have largely been overlooked by the food justice movement, respectability politics offers one potential explanation. Queerness and transness have been disregarded as factors related to just access to food and food production. Further, we ask how the food justice movement can resist idealizations of respectability, which label queerness and transness as unsuitable or inappropriate topics to discuss when confronting issues surrounding food production and consumption. As Minian (2013) demonstrates, although the respectability politics strategy offers organizations a pragmatic approach to gathering support, this practice ultimately undermines justice-based efforts to serve community members' needs. Future research must ask, In what ways are whiteness and white dominance refracted through the food justice movements' attempts to gain advancements for low-income communities and communities of color to the exclusion of LGBTQ+ communities, despite the overlap among these communities?

Queering the Farm for Justice

In this section, we examine intersectional queer ideas for resistance in (1) farm business models: the chosen family farm, (2) farmer and activist retention: LGBTQ+ farmer organizing, and (3) land stewardship models: Indigenous queer approaches.

Figure 2.1
A pride flag flies above the barn at Magnetic Fields Farm in rural Vermont. Photo by Michaela Hoffelmeyer.

Farm Business Models: The Chosen Family Farm

Among LGBTQ+ farmers, scholars find more diverse ways of organizing the farm and household than the heteronormative two-person family farm model, with the chosen family farm being a primary model of organization (Hoffelmeyer 2020). The chosen family farm is a core mechanism to mitigate the anticipated discrimination LGBTQ+ farmers face as well as the white heteronormativity and cisnormativity baked into accessing farming resources. LGBTQ+ communities have long leveraged chosen families rather than blood families due to issues of acceptance, safety, and comfort (Weston 1997). Born out of birth family rejection, the chosen family offers LGBTQ+ people a means of accessing support and safety as well as material benefits such as housing and food. For many, the chosen family is a means of survival in the face of heterosexism and cissexism. In the context of agriculture, the chosen family can serve as an alternative form of agricultural production.

Chosen family farms mean that there is not a heteropatriarchal blueprint for organizing land and farm ownership or labor relations, so this unit tends to promote negotiation and imagination about alternative, more cooperative forms of farm organization. Unlike larger, more intentional communities, chosen family farms are small-scale intentional communities and nontraditional farming models. These models allow LGBTQ+ farmers to curate their own communities, including other LGBTQ+ people, partners, ex-partners, friends, and accepting cisgender heterosexual people (Hoffelmeyer 2020). For example, one version of a chosen family farm utilized individual houses and plots of land in close proximity to another. The individuals included a trans woman, a married same-gender couple, and a heterosexual couple. They shared a tractor and labor but also relied on one another to provide safety and support in a rural community where the LGBTQ+ farmers had experienced harassment (Hoffelmeyer 2020). In a blood family, there is limited opportunity to consent to who is in the relationship. In contrast, chosen family is all about ongoing consent, which, in our view, necessarily makes the chosen family farm a feminist project.

The literature documents how even traditional heteronuclear family farms struggle to pay living wages, and so to maintain farm viability they often exploit their own labor as well as others' (Gray 2014). Here, a queer lens highlights how a true food justice approach would reject any household organization that is predicated on the exploitation of labor as neither just nor sustainable.

Farmer and Activist Retention: LGBTQ+ Farmer Organizing

LGBTQ+ farmers have always existed but have not always been organized. For example, the Lesbian Land Movement and Radical Faerie communities, which began in the US in the 1970s (Sandilands 2002; Hennen 2004), were important precursors to contemporary LGBTQ+ farmer organizations, which have blossomed in the last decade. Organizations like the global but US Midwest–born Queer Farmer Network and Northeast Queer Farmers Alliance help LGBTQ+ farmers find each other, identify safer spaces for rural living and farm work, and advertise gatherings like farmer-to-farmer skill shares and even Farm Pride (Hoffelmeyer 2021; Rise and Root Farm n.d.). LGBTQ+-owned farms like Rock Steady in Millerton, New York, are establishing queer food supply chains to feed underserved LGBTQ+ people (Rock Steady Farm n.d.). The Not Our Farm project shares stories of those who are career farmers but do not own the land they work (Not Our Farm n.d.). By decoupling land ownership with idealized farming goals, many of the farmers featured by Not Our Farm reject the family farm model and the exploitation of farm workers, particularly because these farmers have also experienced abuses on farms across different scales and production practices. While this chapter is focused on the US, it is important to note that LGBTQ+ farmer organizing is emerging around the globe, with, for instance, La Vía Campesina member organizations in Brazil and Europe leading conversations about the roles of gender and sexuality in an agrarian future focused on food sovereignty (European Coordination Via Campesina 2021).

These spaces prioritize several organizing concepts that are particularly queer and applicable to food justice activism: bodily sovereignty, queer ecology, and queer joy. Bodily sovereignty stems from the transgender and queer struggle to define one's own body for oneself, to live safely, and to engage in whatever consensual sexual practices one wishes. Bodily sovereignty demands labor justice, which is necessarily anti-capitalist. Queer ecology encourages us to question how people in power use ideas about nature (e.g., defining what is "natural" and "unnatural") to further their own political goals. Queer ecology also points out that like ecological systems, the most resilient social systems prioritize diversity (Bell et al. 2021). Queer joy is about confronting white heteropatriarchy by centering happiness, fabulousness, and sex. This celebration is important for addressing farmer mental health crises and burnout, increasing retention, and strengthening community resilience in the painful march toward justice.

Land Stewardship Models: Indigenous Queer Approaches

As we suggest that the chosen family farm and LGBTQ+ farmer organizing are crucial to bringing queerness into the food justice moment, we must also reflect on the potential recreation of colonial relations. Drawing inspiration from Liboiron's (2021) examination of solutions to pollution, we ask ourselves and those engaging in food justice, are we assuming that the solution to food justice relies on access to Indigenous land? Land ownership—even if by marginalized groups—can recreate colonial land relations. Among the queer, gay, and lesbian back-to-the-land movements, we must ask, back to whose land? Without examining coloniality within and among agrarian projects, alternatives are likely to recreate forms of violence (see Babb et al. 2023 for a farmers market example of this recreation of violence). As such, a queer approach is necessarily an anti-colonial approach.

One example of an organization taking this approach is the Sogorea Te' Land Trust. Their work entails, "through the practices of rematriation,[2] cultural revitalization, and land restoration, Sogorea Te' calls on Native and non-Native peoples to heal and transform the legacies of colonization, genocide, and patriarchy and to do the work our ancestors and future generations are calling us to do" (Sogorea Te' Land Trust n.d.). Through a webinar entitled "Queer Projects on Indigenous Land" in 2022, the Sogorea Te' Land Trust highlighted several examples of queer-led and queer Indigenous organizations that have confronted issues related to coloniality. The Sogorea Te' Land Trust recommends practices such as developing succession and transfer plans with Indigenous groups, using cultural easements for Indigenous people to be able to access the land, developing alternative economies, rematriating land, and telling the story of the land rather than beginning with the history of occupation by a group or a farmer. For example, through a collaboration with the Northern California Land Trust, Sogorea Te' gained the stewardship rights of a garden in West Oakland, later renaming the site Rammay. After the return of the garden to Indigenous people, the site is now used to cultivate native and medicinal plants, provide food for weekly distribution, and host educational activities. We look to these examples as ways to recognize the rich queer histories of Indigenous groups as well as to confront the potential of queer communities recreating colonial relations.

Further, a queer anti-colonial approach also reexamines labor relations in agriculture. Many farmworkers are Indigenous people, even if they are not Indigenous to the particular lands they are working. In an ethnography of fruit farmworkers, Holmes (2013) discusses how Indigenous farmworkers experience extreme discrimination and mistreatment. As such, we argue that a queer food justice project cannot only focus on farm owners but must pay particular attention to farm workers and to power relations in food production.

Conclusion

In this chapter, we first explored how the collision of heteropatriarchy and white supremacy in US farming has contributed to the food justice movement largely missing queerness and transness as critical parts of the movement. The movement overlooked LGBTQ+ marginalization in agriculture because, for example, it failed to sufficiently question the family farm, heteronormative pathways for farm viability, and respectability politics. We then considered queer strategies for realizing food justice movement goals, including the chosen family farm, LGBTQ+ farmer organizing, and Indigenous queer approaches. Queering farm business models, farmer and activist retention, and land stewardship models offer examples of ways for the movement to move forward in undoing heteropatriarchy and realizing its broader goals.

Common, anti-intersectional notions of food justice as racial justice and queerness as whiteness have pervaded the food justice movement. We have argued that focusing on LGBTQ+ communities in agriculture disrupts these common notions, bringing critical intersections to the movement. Furthermore, we have argued that bringing queerness into food justice should not be seen as a distraction from the movement's racial justice goals but, in actuality, a vehicle for realizing them, not only because these communities overlap but because the strength of solidarity fragments when we use the tools of one oppression to fight the oppression of another.

We have offered several examples of ways to queer food justice, but there are many more important areas for future work. Broadly speaking, we need more research on how relations of gender, sexuality, race, ethnicity, nation, religion, class, and ability structure inequities in the food system. This includes more research on the intersections of heteropatriarchy, systemic racism, and settler colonialism in the family farm model. Despite beginning to document the formation of alternative farm household models such as the chosen family farm, research is needed to understand if and how these models maintain viability, reject exploitation, and offer pathways to justice in the food system. For example, how are queer, Indigenous, and anti-colonial approaches supporting farm laborers? On the other end of the food supply chain, we need more research on LGBTQ+ food insecurity and its relationships with racism, rurality, poverty, homelessness, and workplace discrimination.

We need asset-based research that focuses on queer and trans food resistance strategies. Because of systematic discrimination from farm to plate, some LGBTQ+ people have developed creative, alternative strategies that may contain some solutions to broader food system problems. More research on how understandings of gender, sexuality, and queerness vary across racial, ethnic, geographic, and other contexts helps not only to unpack oppression but also to highlight diverse examples of resistance, success, community, and joy.

We need more research that makes gender and sexual relations in the food system more visible. Concretely, we need to add sexual orientation and more expansive gender identity options to the US Census of Agriculture and other major surveys. We need studies that are supposedly about "other" topics to start naming queerness when it appears in their research. Similarly, we need to talk about the family farm not only as a gendered project but also as a sexualized one. This includes research on how many farmers accomplish or fail to meet their own goals because of heteronormative, monogamous relationship dynamics.

In conclusion, we call on future food justice scholarship to examine if and how queerness and transness are being attended to within the food justice movement and the degree to which respectability politics is rearing its head in today's efforts for a more equitable food system. Building a just food system means not only equitable food and farming for the most oppressed, but using marginalized standpoints and knowledges for inward reflexivity, outward action, and subaltern solidarity.

Notes

1. See Pilgeram et al. (2020) for a detailed description of how the US Census of Agriculture has shifted measures of gender and farmer/producer/operator categories over time.

2. Sogorea Te' Land Trust defines rematriation as "Indigenous women led work to restore sacred relationships between indigenous people and their ancestral land. Honoring our matrilineal societies and lineage's way of tending to the land, in opposition of patriarchal violence and dynamics" (Sogorea Te' Land Trust website, n.d.

References

Babb, Angela, Megan Betz, Sang-hyoun Pahk, and Isis Smith. 2023. "Nostalgia and the Protection of White Supremacy at a Public Farmers' Market." *Gastronomica: The Journal of Food and Culture* 23 (2): 84–94.

Badgett, M. V. Lee, Soon Kyu Choi, and Bianca D. M. Wilson. 2019. "LGBT Poverty in the United States: A Study of Differences Between Sexual Orientation and Gender Identity Groups." UCLA School of Law, The Williams Institute. https://escholarship.org/uc/item/37b617z8.

Bell, Michael Mayerfeld, Loka L. Ashwood, Isaac Sohn Leslie, and Laura Hanson Schlachter, eds. 2021. *An Invitation to Environmental Sociology*. 6th ed. Thousand Oaks, CA: SAGE.

Cooper, Brittany C. 2017. *Beyond Respectability: The Intellectual Thought of Race Women*. Urbana: University of Illinois Press.

Dentzman, Katherine, Ryanne Pilgeram, Paul Lewin, and Kelsey Conley. 2021. "Queer Farmers in the 2017 US Census of Agriculture." *Society & Natural Resources* 34 (2): 227–247.

European Coordination Via Campesina. 2021. "Embracing Rural Diversity: Genders and Sexualities in the Peasant Movement." https://www.eurovia.org/wp-content/uploads/2021/06/MEP_ENG_web.pdf.

Gray, Margaret. 2014. *Labor and the Locavore: The Making of a Comprehensive Food Ethic*. Berkeley: University of California Press.

Hennen, Peter. 2004. "Fae Spirits and Gender Trouble: Resistance and Compliance Among the Radical Faeries." *Journal of Contemporary Ethnography* 33 (5): 499–533. https://doi.org/10.1177/0891241604266986.

Higginbotham, Evelyn Brooks. 1993. *Righteous Discontent*. Cambridge, MA: Harvard University Press.

Hoffelmeyer, Michaela. 2020. "Sexuality and Sustainable Agriculture: Queer Farmers in the Northeastern U.S." Master's thesis, Department of Agricultural Economics, Sociology, and Education, The Pennsylvania State University, State College, PA.

Hoffelmeyer, Michaela. 2021. "Queer Farmers: Sexuality on the Farm." In *Gender and Agriculture Handbook*, edited by Carolyn Sachs, Leif Jensen, Paige Castellanos, and Kathleen Sexsmith, 348–359. New York: Routledge.

Hoffelmeyer, Michaela, Jaclyn Wypler, and Isaac Sohn Leslie. 2023. "Surveying Queer Farmers: How Heteropatriarchy Affects Farm Viability and Farmer Well-Being in U.S. Agriculture." *Journal of Agriculture, Food Systems, and Community Development* 12 (3): 111–125.

Holmes, Seth. 2013. *Fresh Fruit, Broken Bodies: Migrant Farmworkers in the United States*. Berkeley: University of California Press.

Horst, Megan, and Amy Marion. 2019. "Racial, Ethnic and Gender Inequities in Farmland Ownership and Farming in the U.S." *Agriculture and Human Values* 36 (1): 1–16. https://doi.org/10.1007/s10460-018-9883-3.

Leslie, Isaac Sohn. 2017. "Queer Farmers: Sexuality and the Transition to Sustainable Agriculture." *Rural Sociology* 82 (4): 747–771. https://doi.org/10.1111/ruso.12153.

Leslie, Isaac Sohn. 2019. "Queer Farmland: Land Access Strategies for Small-Scale Agriculture." *Society and Natural Resources* 32 (8): 928–946. https://doi.org/10.1080/08941920.2018.1561964.

Leslie, Isaac Sohn, Jessica Carson, and Analena Bruce. 2023. "LGBTQ + Food Insufficiency in New England." *Agriculture and Human Values*. 40: 1039–1054. https://doi.org/10.1007/s10460-022-10403-5.

Leslie, Isaac Sohn, Jaclyn Wypler, and Michael Mayerfeld Bell. 2019. "Relational Agriculture: Gender, Sexuality, and Sustainability in U.S. Farming." *Society & Natural Resources* 32 (8): 853–874. https://www.doi.org/10.1080/08941920.2019.1610626.

Liboiron, Max. 2021. *Pollution Is Colonialism*. Durham, NC: Duke University Press.

McFadden, Ciarán. 2020. "Discrimination Against Transgender Employees and Jobseekers." In *Handbook of Labor, Human Resources and Population Economics*, edited by Klaus F. Zimmermann, 1–14. Cham, Switzerland: Springer International.

Movement Advancement Project. 2019. "Where We Call Home: LGBT People in Rural America." www.lgbtmap.org/rural-lgbt.

Minian, Ana Raquel. 2013. "'Indiscriminate and Shameless Sex': The Strategic Use of Sexuality by the United Farm Workers." *American Quarterly* 65 (1): 63–90. https://www.jstor.org/stable/41809548.

Not Our Farm. n.d. Accessed April 6, 2023. https://notourfarm.org/.

Pilgeram, R., K. Dentzman, P. Lewin, and K. Conley. 2020. "How the USDA Changed the Way Women Farmers are Counted in the Census of Agriculture." *Choices* 35 (1): 1–10. https://www.choicesmagazine.org/choices-magazine/submitted-articles/how-the-usda-changed-the-way-women-farmers-are-counted-in-the-census-of-agriculture.

Rise and Root Farm. n.d. "Welcome." Accessed April 6, 2023. https://www.riseandrootfarm.com/.

Rock Steady Farm. n.d. "Welcome." Accessed April 6, 2023. www.rocksteadyfarm.com.

Rosenberg, Gabriel. 2016. *The 4-H Harvest: Sexuality and the State in Rural America*. Philadelphia: University of Pennsylvania Press.

Sachs, Carolyn E. 1983. *The Invisible Farmers: Women in Agricultural Production*. Totowa, NJ: Rowman & Allanheld.

Sachs, Carolyn E., Mary E. Barbercheck, Kathryn J. Braiser, Nancy Ellen Kiernan, and Anna Rachel Terman. 2016. *The Rise of Women Farmers and Sustainable Agriculture*. Iowa City: University of Iowa Press.

Sandilands, Catriona. 2002. "Lesbian Separatist Communities and the Experience of Nature." *Organization & Environment* 15 (2): 131–163. https://www.doi.org/10.1177/10826602015002002.

Sogorea Te' Land Trust. n.d. Accessed June 13, 2025. https://sogoreate-landtrust.org.

U.S. Department of Agriculture National Agriculture Statistics Services. n.d. "Table 52. Selected Producer Characteristics: 2017 and 2012." Accessed August 31, 2023. https://www.nass.usda.gov/Publications/AgCensus/2017/Full_Report/Volume_1,_Chapter_1_US/st99_1_0052_0052.pdf.

U.S. Department of Agriculture National Agriculture Statistics Services. 2019. "2017 Census of Agriculture Highlights: Farm Producers." www.nass.usda.gov/Publications/Highlights/2019/2017Census_Farm_Producers.pdf.

Watson, Spencer, Oliver NcNeil, and Bruce Broisman. 2021. "The Economic Well-Being of LGBT Adults in the U.S. in 2019." Center for LGBTQ Economic Advancement & Research. https://lgbtq-economics.org/wp-content/uploads/2021/06/The-Economic-Well-Being-of-LGBT-Adults-in-2019-Final.pdf.

Weston, Kath. 1997. *Families We Choose: Lesbians, Gays, Kinship*. 2nd ed. New York: Columbia University Press.

Whatmore, Sarah. 1991. *Farming Women: Gender, Work and Family Enterprise*. London: Springer.

Whitt, Christine, Noah Miller, and Ryan Olver. 2022. "America's Farms and Ranches at a Glance 2022 Edition." U.S. Department of Agriculture Economic Research Service. https://www.ers.usda.gov/publications/pub-details?pubid=105387.

Wypler, Jaclyn. 2019. "Lesbian and Queer Sustainable Farmer Networks in the Midwest." *Society & Natural Resources* 32 (8): 947–964. https://doi.org/10.1080/08941920.2019.1584834.

3 Toward a Decolonial Cuisine: The Entangled Politics of Food Revitalization in Native-Led Culinary Organizations

Ash McLeod and Eleanor Shoreman-Ouimet

In the restaurant industry, to "86" something means to get rid of it, to take it off the menu because it is no longer available. At Owamni, an Indigenous restaurant in Minneapolis, staff t-shirts, as well as the business's social media accounts, are emblazoned with the slogan "#86colonialism." When asked semifrequently what the #86colonialism on the back of our uniform means, the typical response is "To take aspects of colonialism off of the table by offering dishes composed only of precolonial ingredients." That being said, for many team members, 86ing colonialism does not apply only to decolonizing food but also to decolonizing restaurant culture by attempting to remove colonial, capitalist, and often toxic work practices from the kitchen environment.

Over the last decade, discussions about food sovereignty, food (in)security, and narratives about reclaiming "traditional" foodways have become more common in the areas of public health, Indigenous studies, and human rights. Within this same span of time, various Native chefs, growers, organizers, and culinary entrepreneurs have started their own food businesses prioritizing Indigenous knowledge, ingredients, and agricultural methods. Though initially slow to achieve widespread recognition in the mainstream US culinary scene, some of these Native food producers are beginning to garner more attention for their food as well as for their food justice work, resulting in numerous culinary awards. Sean Sherman, an Oglala Lakota "Sioux" chef and owner of Owamni, has won awards from the James Beard and Julia Child foundations and was named one of *Time* magazine's most influential people of 2023.

This chapter highlights the strategies, motivations, and challenges faced by two partnered Indigenous culinary organizations—a culinary education nonprofit and a full-service restaurant—as they strive for food justice and aspire toward a decolonial food system. I[1] first became involved with the leaders of these organizations when volunteering with their catering company prior to the COVID-19 pandemic. As a non-Native anthropologist, I was conscious of cultivating relationships and the groundwork for a collaborative research project before returning for an extended study. In 2021, I began participatory observation, finding employment with the restaurant as a line cook and front-of-house employee and with the nonprofit as a culinary and education associate. My second year of study with them produced thirty interviews with staff members of the organizations. This chapter was written based on my doctoral dissertation, on which interviewees have had the chance to provide feedback. I remain in conversation with research participants, allowing them the chance to read and reflect on how their lives and roles are represented in the work, refine specific arguments with their input, and choose photos and other details collaboratively.

Native culinary organizations exist at a tenuous crossroads, for they occupy political spaces as expressions of Indigenous identity and decolonization (see chapter 5, this volume), and yet the businesses are also subject to colonial capitalist market demands. As these organizations find widespread mainstream success, they also question their own "decolonial" methods. In examining the complexities produced in this emerging arena, this chapter offers new insights into how Indigenous foods and identities are being (re)defined and (re)interpreted by food producers. How these interactions are navigated has implications for the longevity of Native businesses, decolonial food justice, and the ongoing relationships between Indigenous food producers, community members, and their patrons.

Owamni and NĀTIFS

Owamni is a James Beard Award–winning modern Indigenous full-service food establishment that opened in summer 2021. The restaurant utilizes a variety of techniques to "decolonize" plates, offer culinary education opportunities, and provide Indigenous food to the local community. The restaurant presents a variety of dishes inspired by and drawing from the diaspora of Indigenous food traditions all across Turtle Island, from North to South America. They do not serve any foods brought during colonization, including beef, chicken, dairy, wheat flour, and cane sugar. One can expect to find offerings that exemplify seasonality and a respect for broader Indigenous American culinary traditions, such as the craft of making hand-made tortillas or arepas, while pairing those foods with local Indigenous culinary staples such as bison, walleye, and wild rice.

Founded in 2015, North American Traditional Indigenous Food Systems (NĀTIFS) is an Indigenous culinary nonprofit with ambitions of addressing economic and nutritional health inequalities in Native communities by reestablishing Native foodways through culinary education. The nonprofit space functions as a production kitchen as well as a training center for Indigenous entrepreneurs and chefs to learn and grow their cooking or business skills before taking that knowledge and talent back to their own communities. The nonprofit also manages a medicine and edible plant garden and hosts Native school groups, educators, interns, and local entrepreneurs for trainings and workshops. They utilize their garden and other food aid projects to provide Indigenous food and medicine to elders, unhoused relatives, and community activists. NĀTIFS has opened the Indigenous Food Lab Market, which serves as a hub for Native purveyors to sell their food and wares, and they have worked with the USDA to add more healthy and culturally relevant ingredients and recipes to the federal Food Distribution Program on Indian Reservations (USDA 2023).

What Is Decolonial Food?

Complexities and friction arise not only as Indigenous food is defined by Native culinary organizations but also in how they showcase "decolonization" in their cuisines and workplace philosophies. As these organizations find widespread mainstream success, they are both praised and criticized for their specific "decolonial" methods and food sovereignty goals by community members, employees, and patrons.

To contextualize some foundational definitions, food sovereignty entails the right of all peoples to healthy and culturally appropriate food and the right to define their own food and agricultural systems (Desmarais 2007; Figueroa-Helland et al. 2018). Though there has (and will likely continue to be) much debate on the definition of decolonization, scholars Waziyatawin (Wahpetunwan Dakota) and Michael Yellow Bird (Sahnish/Arikara, Hidasta) (2005) define it as "the intelligent, calculated, and active resistance to the forces of colonialism that perpetuate the subjugation and/or exploitation of our minds, bodies, and lands, and it is engaged for the ultimate purpose of overturning the colonial structure and realizing Indigenous liberation" (2). Others, such as Eve Tuck (Unangax̂) and Wayne Yang (2012), remind us that decolonization is not simply a metaphor for social justice, for it specifically requires the repatriation of land back to Native people. When interviewed, several staff members themselves cited Tuck and Yang to contextualize their own decolonial philosophies, explaining that the work we are doing to decolonize foodways, education, and our minds is important, but without reclamation of land rights, it can become little more than a buzzword to be co-opted by white liberals and academics. Speaking about those who would co-opt the term, Riva Garcia (they/them) (White Earth Ojibwe, Dakota, White, Mexican)[2] stated, "[Tuck and Yang] is a very good one, and I want to tell everyone. I want to say 'decolonization is not a metaphor' all the time because I feel like a lot of times, people do see it as a metaphor."

Owamni characterizes itself as a "decolonized restaurant experience," with the core philosophy listed on their website being to "showcase *modern* [emphasis added] Indigenous foods. This means that we prioritize purchasing from Indigenous producers first, and build our seasonal menus to reflect region, culture, and specific Indigenous identities. We have cut out colonial ingredients . . . instead highlighting the true agricultural products of North America, such as corn, beans, squashes, wild game, birds, fish and Native plants. This approach offers a different story of American cooking than the Eurocentric one that is traditionally touted—instead celebrating the amazing Indigenous diversity of Turtle Island" (Owamni 2023).

At times, patrons have been disappointed to not find frybread or Grandma's chicken wild rice soup on the menu. On the other end of the spectrum, the restaurant has also been critiqued for utilizing fine-dining elements and French culinary techniques. Positing Owamni's food as "modern" Indigenous fare has therefore been a strategic decision, for it has allowed staff to have the flexibility of utilizing precolonial ingredients paired with techniques and culinary influences that aren't strictly precolonial or traditional while defining a cuisine all their own.

The different conceptualizations that staff members have for Indigenous food, decolonial food, and a decolonial workplace lend themselves to better understandings of employees' fundamental goals for their organizations. They also help explain why there has been friction within these organizations based on individual decolonial philosophies, particularly those of founders and managers as opposed to hourly employees, such as chefs. Definitions of decolonial food exist on a continuum, with many interviewees defining it as something with completely precolonial ingredients. Others would say that decolonial food can be more flexible or open to interpretation than a strictly precolonial mandate. However, a common thread from staff that distinguished precolonial food or even Indigenous food from decolonial food was *intent*. They described decolonizing one's foodways as a *process* that took a conscious effort or relearning. Eating an ear of corn might be eating a precolonial food, but by learning the stories of the corn, (re)learning how to have a respectful and reciprocal relationship with the corn, or trying to make traditional recipes with corn—according to staff, that would be decolonial. Giving the food and history the respect and time they deserve, reclaiming what was lost or taken, and making a conscious effort to spotlight Native ingredients can all lend themselves to a philosophy of decolonial foods. For example, for Darryl Montana (he/him) (Tohono O'odham), a restaurant line cook lead, a big motivation for wanting to participate in Native food revitalization work was "retaking our fishing rights back, our foraging rights back, and our hunting rights back. With the restaurant . . . we're able to showcase our Turtle Island Indigenous meats, which aren't *exotic* because they were already here before colonization."

Still others acknowledge that, for themselves, despite many cooking techniques and tools having French or colonial origins, "contemporary" decolonial food can still utilize them, for they recognize the right of Native peoples to incorporate outside influences and embrace cultural (ex)change just as every other culture does. The nonprofit's Culinary Program Specialist, Vern Defoe (he/him) (Anishinaabe, enrolled in Red Cliff, WI), explained, "Even other colonized places were able to develop their own cuisine, like Mexican food in Mexico . . . We're still in the eating out of necessity and for nutritional value stage. We didn't really get to develop cuisine outside of frybread really, so I think that . . . we're in the stage where we're kind of creating it, however that comes out, so the easiest way to do it is using colonized techniques." Vern and other chefs are reflecting on what it means to create a contemporary Indigenous cuisine all their own, combining precolonial foods and techniques with other tools and ingredients that might be more familiar and accessible (figure 3.1).

Figure 3.1
Esiban Parent (he/him) (Lac Courte Oreilles Ojibwe), an Owamni server, bartender, and seasonal cook, displaying a plated course from Owamni's winter tasting menu in March 2022. Photo by Ash McLeod.

Why Work in Decolonial Food?

The motivations that employees have for involving themselves in a "decolonial" or "Indigenized" organization are varied, and yet they all speak to their personal philosophies and understandings of what it means to decolonize foodways and strive for food sovereignty.

Education and Access to Healthy Indigenous Food

Sean Sherman (he/him) (Oglala Lakota), chef-owner of Owamni and a founder of the NĀTIFS nonprofit, maintains, "My vision of food relief was just kind of unapologetically putting healthy Indigenous food out there, even though it might not have been the most popular choice for everybody."[3] Those that cite health as a primary drive for becoming involved with Indigenous food revitalization have been motivated in part by memories of their childhoods when the only foods available were from the commodity foods program or processed snacks from gas stations. Several employees had become prediabetic or had family members pass away from type II diabetes and were excited to work with healthy Native foods as dietary alternatives. Educating customers about the history and importance of Indigenous foods and providing culinary education for aspiring Native chefs and community members is a key goal of both Owamni and NĀTIFS. Joe Wade (he/him) (Bad River), a lead cook at the restaurant, says, "There's a lot of people I grew up with in the culinary scene themselves up there [in Bad River], but no one's really doing food sovereignty, so whatever knowledge I can take from here and share with them, that's really what I'm looking to do."

Yet, despite frequent mentions of revitalizing "healthy" Indigenous foods as part of personal and organizational philosophies countering diet-related health disparities in Native communities, multiple staff members also engage in a more nuanced conversation surrounding Indigenous food, one that is directed at not allowing a language of fatphobia or shaming for choosing non-Indigenous foods to flourish. Rivianna Zeller (they/he) (mixed race Latina), a former restaurant cook turned farmer and distribution coordinator for a Native-led farm, strongly believes that "food has no morality. And it especially gets kind of sticky when you're dealing with Indigenous foods and reconnecting folks in that way because people can start to attach morality to it like, 'Well, you're eating commodity food, and I'm eating organic

Indigenous produce, so who's *really* reconnecting with the ancestors?' That's not what this is about. At all."

What "Indigenous" or "Native" food is according to participants ranges from a completely precolonial diet (i.e., food from before European settlement) to more flexible definitions that allow for culturally significant foods made by Native people from colonial ingredients, such as frybread, or even foods such as Hot Cheetos that are not made by Native people yet are still a common and beloved snack in Native communities. Ma'iingan Sherritt-Stone (they/them) (White Earth Ojibwe), a prep and line cook, embraces thinking critically about the complexity of incorporating both precolonial foods and colonial ingredients into an understanding of Native cuisine: "Wild rice being harvested the traditional way is Indigenous food the same way that fucking frybread and a lot of commodity foods are also Indigenous foods. I think all those things are incorporated into our history and our survival, and I think that's really the basis for what makes Indigenous food a lot of the time . . . I think that definition, the circumstances, and the resources have changed, so I think that also changes the way that we should look at [Indigenous] food and the way that food is made." These sentiments from Rivianna and Ma'iingan demonstrate a critique of the ways that "healthy" or "organic" foods can be uncritically valorized in food movements, and they show a dedication to creating a cuisine that represents ongoing experiences of oppression and community ingenuity while also celebrating a precolonial past.

Being a Good Relative

Another motivation for many staff members is simply to be a good relative by feeding people in need. Being a good relative is baked into the Dakhóta language. Liz Cates (they/them) (Bdewákhaŋtuŋwaŋ Dakhóta), the nonprofit's education coordinator, tells us that "the tenet of being a good Dakhóta person is being a good relative, and a big thing that comes up a lot is *'Mitákuye Owás'iŋ,'* which means 'all my relatives,' and that means every single thing on this earth." Part of being a good relative for so many who work here is to make sure that unhoused relatives and others experiencing food insecurity are fed, or that those on the frontlines of social justice initiatives, such as water protectors, Dakota Access Pipeline activists, or Minneapolis Uprising protesters, are supported with food. As Liz Cate's words indicate, being a good relative also includes showing that same respect to the plant and animal relatives "that give us life and medicine and food and warmth." This may manifest itself as sustainably sourcing meat and hand-harvested wild rice, taking care to cook and serve culturally significant ingredients thoughtfully and respectfully, or leaving tobacco as a thank you to the land when we tap for maple or forage Indigenous plants.

A substantial number of staff members interviewed were motivated to apply as a way to reconnect with a community and food culture that they previously felt distanced from. These perceived disconnections are rooted in a number of sources, such as cultural separation due to assimilation and residential schools, growing up away from ancestral homelands or tribal communities, or a lack of confidence in identifying as a Native person because of colonially imposed blood quantums. Francesca Garcia (she/her) (Mexican, Tarahumara, Yaqui), the Indigenous Food Lab Market's herbal specialist, recounts, "We ended up having to relocate to Southern California from Northern California where we're from, so that kind of separated us from our Indigenous community up there. And it's been kind of hard finding that sense of community again. I think that's something my sister and my mom and I have always been yearning for—to have that back and really try to find our roots and our connections." A number of staff members expressed feeling more confident in their identities and cultural knowledge as a result of working with the two organizations and reconnecting with Indigenous communities through food.

For a majority of the restaurant's interviewed staff, being a good relative also extends to the restaurant industry and workplace culture itself, for staff have expressed a motivation to work there that involves changing the toxic and oftentimes authoritarian, sexist, and racist nature of restaurant culture from within. They desire a space that is safe and equitable for all employees that also diminishes colonial capitalist power structures such as the French brigade system. As Isaac Verley (they/them) (Fon du Lac Band of Ojibwe), a line cook, discussed: "The toxic masculinity that goes on in the brigade structure is what everybody who came [here] was trying to get away from, because all fine-dining restaurants do that. We didn't want to have animosity between coworkers." Some managers have also expressed the motivation to do more for their staff surrounding uniquely Native issues, such as providing paid time off or bereavement leave to attend ceremony, powwows, funeral fires, or sweat lodge.

Friction Due to Capitalist Demands

Perhaps not surprisingly, tensions tend to emerge in Native entrepreneurial spaces as they gain public success, for producer's social justice commitments and motives are put on display and called into question. Food sovereignty projects have thus become environments ripe with "friction," which, according to anthropologist Anna Tsing (2005), characterizes the conflicting encounters and complex social interactions produced by diverse cultures coming together in our contemporary globalized world. For example, if Native producers fully embrace the larger capitalistic entrepreneurship model for their businesses, they may be admonished for adopting a colonial model that does little to support the community and other disenfranchised minorities. Yet if they try to shy away from the capitalist model completely, they may risk failed business ventures and a short-lived role in the food justice movement.

While a few team members believe that decolonization of the workplace is possible, if not already achieved, others believe having a decolonized business is completely impossible in our current world. The majority of those interviewed fell somewhere in the middle ground, holding that while a completely decolonial business is not possible, their organizations could still be engaged in a process of self-reflection guided by Indigenous ways of knowing and being. Ismael Popoca Aguilar (he/him) (Mestizo), the nonprofit's culinary program manager, reflects on the possibility of complete organizational decolonization: "I don't think it's a hundred percent decolonized, and that's okay. We're working towards it, and it would be great if we could make history as the first decolonized work environment and do things differently . . . I would hate for someone to come and tell me 'This is how you decolonize a work environment. This is the only way to achieve decolonization.' That is falling into the fault of colonizing." As a result of having "decolonial" organizations that are still very much engaged in a colonial and capitalist system, nearly all of those interviewed acknowledged the challenges of decolonizing a *business*. Both the restaurant and the nonprofit struggle, for instance, to make sure that staff are paid living wages, to reduce reliance on workplace hierarchies, and to deal with the expectations of patrons or donors.

Liz Cates recognized the difficulty of having an accessible decolonial *for-profit* restaurant:

> I'm the first to admit that I think it's great that [the restaurant] exists. . . . It's putting Native foods on the map, and people are so excited to go try it, and . . . that could make a big difference on how someone votes on something. . . . But in the step of greater decolonization . . . there's still work to be done. When houseless people can walk in and eat there too or everyone is getting paid the same. . . . The people that work [at the restaurant] and the people that work [at the non-profit] aren't paid enough to be in the same tax bracket that gets to go [to the restaurant].

A recurring issue that staff members are finding difficult to decolonize is the hierarchical structure found in restaurants and even in nonprofit organizations. As market manager Jason Garcia (they/them) (Latinx and white) explains, "We all grew up in a capitalist society, and we all grew up in a colonial society, so as much as we want to kind of try to retrain our brains, when things get emotionally escalated, we tend to revert back to our early memories and what is sort of instinctual behavior for us, and that can often be to return to hierarchical scenarios and say 'Well, I'm in charge, so therefore, you have to listen to me.'" The nonprofit's former NĀTIFS communications manager Em Matson (they/she/he) (Sault Ste Marie Chippewa/Ojibwe) adds that "it's hard to be a community and have someone who can fire you, you know? . . . We can't just be like 'oh, there's tons of Native people in this organization, so we're decolonized.'"

Another challenge for the restaurant and nonprofit employees involves making concessions or compromises to ensure that patrons and donors are happy. Em Matson further notes that "white donors will donate a lot of money to us, and in that, we end up sometimes having to then look good for white donors, and I think there are compromises that can be made to, rather than look good for white donors, have white donors understand the work that we do is important." Others emphasized the need to strengthen relationships that might have broken between the organizations and the local Native community. Riva Garcia holds that "when we talk about decolonizing a workplace . . . we're a non-profit, so we need to be accountable not just to our funders but to the community we're saying we serve."

As shown, the vast majority of staff members are incredibly passionate about the mission of the restaurant and nonprofit, desiring to be a part of Indigenous food revitalization and food justice, whether that be helping a nonprofit work toward local Indigenous food sovereignty and fighting food insecurity or using the for-profit restaurant to share and educate customers about Native foods. However, because of the challenges and concessions that working within a capitalist and colonial system necessitates, one of the most common points of friction for team members was summarized by Rivianna Zeller: "Using pre-colonial ingredients but still operating in this really intense kitchen system. . . . There's a general disconnect between forward facing values of the restaurant and the internal culture . . . especially in employee treatment and the hierarchy."[4]

Playing the Game Versus Gaming the System

For the producers I have worked with, pushing back against areas of friction involves balancing, in the words of sociologists Pierre Bourdieu and Loïc Wacquant (1992), "playing the game," and what I am referring to as "gaming the system." I use "playing the game" here to describe examples of how employees of these organizations negotiate entanglements with capitalist and colonial systems by making compromises to ensure that their businesses stay open, staff are paid, and the mission of food revitalization continues. Playing the game, therefore, includes paying taxes, complying with FDA regulations and food safety standards, pivoting to meet the expectations of nonprofit donors or non-Native patrons, and generally making strategic business decisions to ensure profit. In contrast, I interpret "gaming the system" as ways that individual staff members or the organizations as a whole are finding to subtly resist capitalist systems or organizational policies that they believe do not exhibit decolonial philosophies. This includes setting work boundaries for personal health, finding ways to use company resources to serve the community without explicit permission from company administrators, burning sage and smudging despite building fire safety policies, pushing leadership for increased wages, and prioritizing Native customers.

In regard to burning sage, or "smudging," the nonprofit's operations manager, Alexis McKinnis (she/her) (White Earth Ojibwe lineal descendent), says, "[Building managers] do not want to get into a position where they are telling an Indigenous organization that they can't do ceremonial practices in their space as long as it's not a health and safety risk, because it's not. And I said 'We are smudging anywhere we goddamn well please,' and that's fine."

Several employees have also taken the important role of feeding community members into their own hands. Isaac Verley recounted a personal strategy used while working for the restaurant's catering company, bringing to-go containers to give away extra food to those in need after events. "We don't advertise! But yeah, I'm the one that always makes sure the extra food actually gets eaten by people." A few mid-tier managers and hourly employees also spearheaded informal projects to go out into the community to provide frybread and chili for unhoused relatives on Indigenous Peoples' Day 2022 as well as cooking for monthly events at Little Earth, an urban Indigenous housing community. Em Matson states, "Helping the everyday person, not just the traps of money . . . I think that's a good way to fight the capitalism aspect of it is 'are we getting Indigenous food into Indigenous people's hands?'" By "gaming the system," staff employing these strategies help the restaurant and nonprofit to stay closer to their decolonial values.

Conclusion

With increased public interest in decolonization, alternative food movements, and award-winning Indigenous restaurants, the mainstream US food scene, which may have once thought of Indigenous food as a cuisine long dead and forgotten, is being exposed to new understandings of Native food and Indigeneity. Native food businesses are becoming arenas where identities are entangled with food culture as these organizations are (re)defining and (re)interpreting what Native food means for themselves and the local community. Decolonization here has become an intentional process, an ongoing goal that has not yet been accomplished, but one that employees are committed to seeing through.

Friction due to the demands of capitalism has become a key challenge for the restaurant and nonprofit, and how alternative food movements and organizations navigate this tension has implications for the longevity of Native businesses, decolonial food justice, and the ongoing relationships among Indigenous food organizations, their patrons, community members, and staff. Despite the friction that has manifested, the majority of team members at both organizations care deeply about the work and each other, believing that the continued effort is worth the internal and external conflict. The passionate chefs, entrepreneurs, organizers, and educators of these organizations are doing everything they can to further food justice and decolonization while Native food is garnering such popularity. Native food producers in these spaces are making it known that, despite friction and colonial forces working against them, by revitalizing Native foods and fighting to decolonize their own food systems, Indigenous food and Indigenous people are here to stay.

Notes

1. Although Eleanor Shoreman-Ouimet is listed as a coauthor, Ash McLeod conducted the ethnographic fieldwork and interviews used in the creation of this chapter. Therefore, "I" statements, rather than "we" statements, were used throughout the chapter to reflect Ash's personal research and analyses.

2. The community affiliations of interviewees are documented throughout this chapter. You may note that I do not always utilize the official government names of Native nations. You may also note that in some cases, I designate a tribe or nation in two different ways for different people, such as one person being identified with Bad River versus a Bad River Band of Ojibwe lineal descendent. This is intentional, as I am directly reflecting how those individuals wish to be identified. Similarly, some participants chose to only include one tribal affiliation as an identifier and others chose to honor all parts of their identities by including multiple tribes as well as some designations like white or Mexican. Some staff members also chose to designate themselves as either enrolled members or lineal descendants of particular nations while others chose not to disclose this information. The specific ways and care with which participants chose to identify themselves should be respected.

3. Though Sean Sherman is a current face of Indigenous food revitalization and a driving force behind Owamni and NĀTIFS, this chapter is not meant to represent the voice and works of Sean alone. Rather, in the spirit of Indigenous food sovereignty, which has and continues to be a community-based endeavor, this chapter speaks to the varying perspectives of a collective of individuals who have made this project, and these organizations, possible.

4. A number of respondents acknowledged that hierarchies existed in precolonial Indigenous societies and that hierarchies are not inherently harmful or unjust. Rather, the hierarchies they spoke of changing or dismantling in this chapter were those that they specifically felt misrepresented their interests or contributed to the toxicity of restaurant work culture.

References

Bourdieu, Pierre, and Loïc J. D. Wacquant. 1992. *An Invitation to Reflexive Sociology*. Chicago: University of Chicago Press.

Desmarais, Annette Aurélie. 2007. *La Vía Campesina: Globalization and the Power of Peasants*. Halifax, Canada: Fernwood.

Figueroa-Helland, Leonardo, Cassidy Thomas, and Aguilera Abigail Pérez. 2018. "Decolonizing Food Systems: Food Sovereignty, Indigenous Revitalization, and Agroecology as Counter-Hegemonic Movements." *Perspectives on Global Development and Technology* 17 (1–2): 173–173. https://doi.org/10.1163/15691497-12341473.

Owamni. 2023. "Owamni." NĀTIFS. Accessed November 30, 2023. https://owamni.com/.

Tsing, Anna Lowenhaupt. 2005. *Friction: An Ethnography of Global Connection*. Princeton, NJ: Princeton University Press.

Tuck, Eve, and K. Wayne Yang. 2012. "Decolonization Is Not a Metaphor." *Decolonization: Indigeneity, Education & Society* 1 (1): 1–40.

USDA (US Department of Agriculture). 2023. "USDA Indigenous Food Sovereignty Initiative." USDA. https://www.usda.gov/tribalrelations/usda-programs-and-services/usda-indigenous-food-sovereignty-initiative.

Waziyatawin, Angela Wilson, and Michael Yellow Bird. 2005. *For Indigenous Eyes Only: A Decolonization Handbook*. School of American Research Native America Series. Santa Fe, NM: School of American Research.

4 Plotting Our Legacy: Radical Black Ecology from the Provisional Plot to the Garden Lot

Alexis Wiley and Akilah Chatman

Introduction

For Black people throughout American history, our expected and realized relationality to nature has been defined by our history of labor on the land through slavery and our dispossession from land in the aftermath of it. Our ancestors were used to enact settler colonialism by cultivating "nature," which plantation owners understood as wild and untouched despite the presence of Indigenous peoples, into so-called civilized land (McKittrick 2013, 6) that supported the plantation class's control, surveillance, and terrorism of enslaved Africans. In *Plantation Futures,* Katherine McKittrick (2013, 3) describes the plantation structure as the foundation of the city; the city today perpetuates parallel violence against Black people as the plantation did during slavery. White colonial plantocracies rendered enslaved Africans and their descendants "fungible" (King 2019). In agriculture, fungibility means that seeds are seen as a disposable commodity that can easily be replaced by one another, and thus, it does not matter if some of them fail to grow. Plantation agriculture reduced nature to the commodity of land and Black people to the commodity of laborers—a departure from our previous relationships with nature that manifested in foodways, religions, and overall culture. Yet despite the severance and adulteration of land-based relationships under isolating slavery regimes, Black people continue to develop methods of resistance to retain their relationship to land and culture.

The ecology of the plantation that rendered Black people fungible simultaneously created the conditions for Black fugitive formations, "creative practices of refusal, nimble and strategic practices that undermine the category of the dominant" (Campt 2014, 87; see also Simpson 2016). In this chapter, we trace these formations in heritage farming from the provisional plots of the plantation to present-day urban community garden lots. These past and present spaces of fugitivity generate ecosystems, interconnected geographic and temporal areas that maintain themselves through their interdependent relationships and enable Black people to renegotiate the terms of their exploitation. In both ecosystems, Black people are stepping into a liminal space where and when we embody the agricultural experience of our ancestors, making new geographies that reaffirm our senses of place and being. In this chapter, we germinate the term "ontological succession," drawing on processes of ecological change to describe the mechanism by which ontologies from past geographies, the plot and the plantation, have progressed to present geographies of the lot and the city. Ontological succession is the path by which ancestral practices manifest in fugitive practices of today. As such, we have written this chapter in reverse chronology to acknowledge the atemporal ontologies of the plot and the plantation, and their disorienting effect. To unearth these "secretive histories" (Wynter 1971) of what we call the plot and the lot, we use radical Black ecology, defined as "Black thinkers, movements, communities that have refused the ruse that capitalism, the state, heteropatriarchy, and the domination of more-than-human nature are the means and ends of justice and freedom" (Opperman 2020).

Fungibility manifests on the plantation and in the city where the ecology of these sites marks Black people as less than human, eliminating Black people's agency and self-determination. Fugitivity functions as the interval where Black people find ourselves between enslavement and freedom within "the paradoxical space" (McKittrick 2006) of the plantation or the city (Hartman and Best 2005, 3). Throughout the historical process of slavery and what Saidiya Hartman calls its "afterlifes" (2007), there is a constant interplay between fungibility and fugitivity. Just as Black people developed the plot on the plantation, we developed the garden lot in the city. Here, the fugitive work on the plot and the lot subverts the conditions of fungibility dictated by the plantation and the city. As the sites of fungibility render Blackness as nonhuman, the projects of fugitivity expand Blackness to encompass our creative geographic formations, reclaiming our self-determination and relationships to the land. Despite the fungible conditions generated by the plantation's logic, Black people have developed spaces of fugitivity in the past and present—what we call the plot and the lot—through which we assert our agency by returning to the right relationship with the land and reclaiming our foodways.

Before presenting our work, we acknowledge that our experiences as Black, queer environmental stewards based in Philadelphia, Pennsylvania, influence this chapter. As a heritage farmer and ecologist (Lex) and an environmental justice researcher (Akilah) engaged in our local community agriculture scene for the past five years, we have lived the experiences discussed in this chapter. Our relationships to food and environmental justice work in our community proved to be critical tools to theorize through this text.

The Plantation and the City as Geographies of Fungibility

The City as a Modern Fungible Ecosystem

We identify the city as within "a plantation logic . . . characteristic of slavery [that] emerges in the present both ideological[ly] and materially" (McKittrick 2013, 3). Humans were extracted from various parts of the African world to cultivate agricultural commodities across the Americas to form the plantation, developing a world of exchange value and enabling white survival and the conquest of Indigenous American land (Carney and Rosomoff 2009).

McKittrick (2013, 3) summarizes George Beckford's "plantation thesis," which describes how plantations' systems generated city wealth: "This plantation history not only generated North Atlantic metropolitan wealth and exacerbated dispossession among the unfree and indentured. It also instituted an incongruous racialized economy that lingered long after emancipated and independence movements in the Americas; and that the protracted colonial logic of the plantation came to define many aspects of post slave life."

Christina Sharpe (2016, 3) describes how this plantation logic constitutes "the wake" of slavery that serves as a framework of and for Blackness in the African diaspora that is still unfolding in the aftermath of Atlantic chattel slavery. It is important to state explicitly, then, that the anachronistic nature of the plantation has woven itself into the modern formations of our society. Slavery is not only a past phenomenon—it is our present.

Today, Black people are dispossessed of our free agency in the areas in which we reside. The slums we are pushed to are heavily polluted, overpoliced, and strategically lacking access to healthy food—otherwise known as food apartheid (Taylor 2014; see chapter 6). Food apartheid is not solely an accessibility issue. It is an attack on the role food plays in cultural identity, community cohesion, place-making, and self-sufficiency (Washington n.d.; Garth and Reese 2020). We also see the disruption of Black people's agency emerge from the health issues caused by environmental pollution and the disruption of community cohesion from the unjust incarceration that attacks the ability of these communities to be sovereign (Gilmore 2007; see chapter 20).

Yesterday's plantations and today's cities work to control Black livelihood. Capitalism continues to place a cost on the life and lived experiences of Black people, just as it was designed to. The city has been configured to continuously recreate fungible ways of being for Black people. But just as in the plantation, in the city, we witness the conflation of life with labor yielding to the cultivation of new forms of being. Though the parts of the city Black families were forced to and continue to inhabit have turned decrepit, they are also microhabitats where fugitive ways of being are continuously founded. In the city, we find new ancestral connections, new ecosystems emerging, and a rejection of the "demoralization of the slums" (Hare 1970, 4). Where the "city" has created food apartheid, the inhabitants of the "slums" have called upon their ancestors to create modern plots in the form of community gardens.

The Great Migration: How the City Succeeds the Plantation

After emancipation, the nation went through a series of sociopolitical transitions attempting to rectify centuries of slave labor by economically supporting newly emancipated Black folks through policies like Special Order No. 15, which awarded "40 acres and a mule." This, of course, resulted in both government-sponsored and quotidian backlash from white communities, including the development of Black Codes, vagrancy laws, Jim Crow laws, disenfranchisement, lynching, and the founding of white supremacist organizations, to name a few methods (Wilkerson 2010). Although Black communities began organizing around land tenure (White 2018, 28), agriculture (, Gilbert and Eli 2000, White 2018), voting rights, fair wages, and labor laws, millions of Black people fled the South during The Great Migration (1910–1970) in the wake of this resurgent terrorism (Gilbert and Eli 2000, 68) that sought to maintain their social position as fungible.

However, they would still find themselves interlocked in the same racist dynamics they fled. In addition to the estimated 8 million Black Southerners who migrated out of the South, there are an estimated 20 million white Southerners who moved alongside them (Gilbert and Eli 2000, 68), albeit for different reasons. Like the Black families carrying their heritage through traditions, seeds, stories, and other things, white migrants brought what they imagined to be theirs—white supremacy—contributing to the already "covert" racism of Northern and Western cities. Segregation, work discrimination, criminalization, disenfranchisement, disrupted land tenure, lynching, racial massacres, and other factors haunted the very Black families that fled the South to escape these practices. In this way, the city succeeds the plantation ontology.

Plantations as Sites of Fungibility

American plantations were located in semitropical and tropical zones of the continents, including southeastern and Gulf coastal states of the US, the Caribbean islands, Central America, and South America. Enslaved Africans exploited by European colonizers transformed vast areas of natural land into plantations of crops such as cotton, rice, and sugarcane. Chattel slavery, enacted in the sixteenth century, legalized the ownership of the enslaved as "property" that could be bought, sold, and owned forever, including their children—generations of enslaved folk creating generational wealth for white people. The wealth generation of European colonial plantation economies was bolstered by their exploitation of the knowledge and bodies of enslaved Africans through the strategic human trafficking from the West, Central, and Southeastern coasts of the continent where some of these cash crop cultivars originate (Carney and Rosomoff 2009, 76; Harris 2012; Twitty 2018). Enslaved Africans were exploited not only for the cultivation of cash crops but also for provisional food production in the early days of the colonies, particularly in the tropics, where European food varieties were difficult to grow. Without the knowledge, skills, and labor force of enslaved Africans, colonists would have starved (Carney and Rosomoff 2009, 105).

Chattel slavery created a pervasive geographical fixture where, from the vantage point of white people, the social order was clear, with white men at the top and Black people at the bottom. Enslaved people were simultaneously stripped of their self-determination and stripped of their sense of place as their hair was cut, their names changed, and their native languages, religion, and foodways forbidden. Within this plantation logic, the bodies of the enslaved were considered always available, always willing for work, sexual exploitation, entertainment. and many other forms of violence (Story 2015, 26–27), rendering them fungible.

The plantation system was designed to eliminate the agency of enslaved Africans as they were dominated by the external forces perpetuated by slaveholders. Within the space-time of the plantation, Black people resisted in a plethora of ways, including in their provisional garden plots. In these plots they grew crops uncommodified by the plantation, gathered in community, and traded among themselves, engaging in the "roots of culture" in the plantation that "was the superstructure of civilization" (Wynter 1971, 6).

The Plot and the Lot as Geographies of Fugitivity

The Lot as a Modern Fugitive Ecosystem

The lot is a vacated space brimming with potential that, as we explore here, Black communities cultivate into new ecosystems of being. We, as authors of this chapter, formed this ecosystem model from our experiences working, volunteering, and educating with primarily Black and Indigenous American community gardens in Philadelphia over the past five years. Our experiences as farmers grant us the opportunity to witness and embody this place-making work firsthand, and we reference our two deepest connections as examples.

Sankofa Community Farm at Bartram's Garden in Southwest Philadelphia prides itself on being "an intergenerational, spiritually-rooted, African-diasporic" farm. Thirteen years ago, the cofounders, Chris Bolden Newsome and Ty Holmberg, and early youth cohorts of the farm's program transformed an abandoned baseball field into a flourishing site for food sovereignty. Now, farmers, community gardeners, youths, elders, volunteers, and other community members continue to engage in Afro-diasporic natural agriculture traditions like planting in soil mounds and intercropping, tending to the land and growing food like our ancestors did. Much of the food produced is a part of our "Sankofa Signature Crops," which highlight crops endemic and adapted to the foodways of the African diaspora. The staff features the signature crops in our education and culinary efforts to provide affordable access for the neighbors that recognize these foods and to "re-member" our community's relationships. Sankofa recognizes that our community lost access to and knowledge of our traditional foods through waves of displacement, so our primary mission is to reroot the Black community to our larger cultural, ecological, and culinary practices. Though it is an African diasporic space, the farm opens itself to everyone, understanding how the ways of being and engagement in this space are a model for how all communities can create for their own, evoking the praxis of radical Black ecology.

On the other side of the city in Kensington, the Norris Square Neighborhood Project has a similar history of converting abandoned lots into sites of community and cultural engagement.

Deindustrialization and rising drug use in the 1970s destabilized neighborhoods all over Philadelphia, as many essential businesses closed. After an antidrug raid in the 1980s, conditions worsened in the Kensington neighborhood. Cofounder Iris Brown and other Puerto Rican women residents of the neighborhood formed Grupo Motivos and organized to provide support for their community. Working with the founder of Norris Square Neighborhood Projects, Natalie Kempner, the women founded community gardens in the neighborhood (Bartlett 2022). Today, these gardens honor the jíbaro, Taíno, and African roots of Puerto Rican heritage by featuring crops, agricultural techniques, building structures, costumes, artwork, kitchens, and much more to instill pride in the youth and neighbors they serve (Bartlett 2022).

Though we only explore two examples here, community gardeners across the city grow and gather, generating their sovereignty through the gardens, regardless of whatever other entities would like to deem themselves involved. In Philadelphia, 56 percent of these sites are founded on land-insecure vacant lots owned by the city or private investors with no invested care (City of Philadelphia Parks and Recreation 2023, 10), so neighbors take back the lots. Despite the lack of land tenure, gardeners organize to design, build, plant, cultivate, and gather in their self-determined spaces. It is an ecosystem where Black folks continue to engage in their freedom-founding legacy, creating their own society with their own rules.

Rather than practicing in the capitalistic exchange value market, the gardens thrive on the use-value market, where the community's self-identified needs determine what is produced. The community gardeners grow the foods that nourish the bodies and souls of the people who connect to them, valuing their health physically, mentally, and emotionally.

The work on the lot undoes the notion that Black relationality to nature is only through labor. Here, "reawakening ancestral connections to the land is indeed possible in the city" (Strauss 2019, 81). On the lot, growers revive and preserve the traditions of previous generations that have been lost to us through the generational trauma of transatlantic slavery. Many grow heritage crops like watermelon, okra, and black-eyed peas that are endemic to Africa or naturalized to Black foodways, using ancestral farming techniques that rematriate their relationship with the land. It is food justice and sovereignty for us to grow our cultural foods with culturally relevant methods. The lot allows us to shift our relationship with the land and food production from one of labor to one of connection. We rediscover that we are not separate from the land; rather, we shape the land and the land shapes us. By that same token, the food we grow not only nourishes our bodies but also stewards deeper connections to our culture.

This practice of farming through ancestral connections to nourish the soul, also known as heritage farming, is political. On the lot, we unsubscribe from the rules of the city. Instead of fighting to be valued in this system founded on Black labor, we can lean on our ancestors to create a system that is for us. Heritage farming is a path forward. It creates a new ecosystem of being centered on the experience of the community, where the land is no longer separate but is a member of the community. While the city continues to replicate the trauma of the plantation, heritage farming roots our resistance in the everyday resistance of our ancestors' practices on the plot.

Birth of the Garden Lot: How the Lot Succeeds the Plot

These community gardens on vacant lots served as sites of connection for Black migrants displaced during The Great Migration and international immigrants fleeing similar conditions (Strauss 2019, 81). Many urban community gardens began as victory gardens during the world wars, which garnered support from city governments (see chapter 15). However, following rapid deindustrialization in the 1970s and 1980s, many industries abandoned their property, leaving behind vacant lots. "White flight" accelerated as many white neighbors moved into suburban regions adjacent to the city, furthering turnover to vacant lots. Coupled with diminishing populations and the rise of the crack and opioid epidemics, Black and immigrant neighbors began to reclaim lots that served as sites for drug sales and cultivated them into gardens.

In these lots, much like the plots on the plantation, Black folks are engaging in a form of ontological succession by cultivating sites of fugitivity by engaging their ancestral knowledge and relationship to the land through agriculture. On these lots situated in urban landscapes that, like the plantation, render Black bodies as "labor" and "disposable," Black folks subvert that geography. Choosing to grow ancestral foods, building gathering spaces for education and economic sovereignty in intergenerational communities on collectively owned land on our terms, often without permission, creates an entirely new ecosystem. The ecology is organized through relationships to one another, to nature, and to the ancestors which succeeds the ontology our ancestors created on the plot. Though the city inherited the ontology of the plantation, succeeding in the project of subjugation, as always, Black folks have interrupted it by holding on to our "own ontologies."

Like their families down home, many Black families found themselves relegated to the poorest lands of the cities. And still, they continued to organize for land ownership, education, voting, employment, and other rights, demanding more from a system that never intended to include them. They made the most of what they did have, calling on a longer-standing legacy and turning to the land to remain rooted in themselves.

From our experiences and our research in the archives, we know that such a large movement of people, especially Black folks who fought to keep much of their African heritage during the displacement of slavery, undoubtedly means that our culture moved with us. Our people brought recipes, stories, seeds, and skills to these geographies. Thus, it is not only our migration, but the succession of our place-based knowledge that allows us to rework the ontology of the plot into the contemporary lot.

Plots as Sites of Fugitivity

Not every plantation granted the enslaved a site to grow subsistence crops. However, for those that did, three methods emerged: individual plots, provision grounds, and yards around their cabins (Carney and Rosomoff 2009, 127–138). For our purposes, we are collectively naming these sites the plot.

Though not organized as rebellions or flight from plantations, the plot was a quotidian act of resistance. It is the geographical site where Black people unworked the nastiness of the plantation into geographies that redefine what it means to be human at this time and what their humanity can look like. Using Tiffany Lethabo King's metaphor of porosity, we agree with their assertion that "as fungible bodies, Black bodies are porous and can merge with . . . nature [and their] connections with plants and nonhuman forms of life confound Enlightenment ordering of the world" (King 2019, 130). Within the plot, we observe these dynamics as enslaved Africans cultivate crops endemic to their places of origin and incorporate crops from the New World to support their holistic survival, honoring relationships with nature that value connection over domination.

For Katherine McKittrick, plantation geographies are about racialized captivity in which the plot serves as "both a site of self-captivity and a loophole of retreat" (McKittrick 2006, 73). In both contexts, the "loophole" creates a sense of place where the enslaved can explore their opportunities within carceral geographies. The access to land for independent food production was an opportunity for the enslaved to grow food without supervision, supplementing their diets and producing foods beyond their needs to circulate among their communities. In these plots, enslaved Black people grew indigenous African crops that arrived in the Americas with them as provisions for the slave ships. Coupled with the introduction of newly arrived Africans to these colonies during the transatlantic trade, enslaved people not only had the crops but also the additional knowledge of African agricultural practices for cultivation. Wynter (1971, 5) describes that they transplanted their values from their ancestral African societies such that "the land remained the Earth—and the Earth was a goddess; man used the land to feed himself; and to offer first fruits to the Earth."

The plot became more than a space to grow food. It became a fugitive ontology, a cultural ecosystem, and "a site of folk culture" (Wynter 1971). Enslaved people's ability to grow their ancestral crops or substitute for key ingredients allowed them to continue their cultural relationships with nature and embody their ancestral traditions, offering spiritual respite. Growing for themselves away from the watchful eye of the plantation undoubtedly offered an opportunity for pleasure where the enslaved could engage among themselves. In these spaces, the enslaved created new geographies that centered relations against the fixed dynamics of the plantation ecosystem. These formations are the origins of the engagement in food justice in the plot and the lot.

In these plots, Black people subverted the colonial plantation narrative that they only had their bodies to offer for labor. The fugitive work they accomplished on the plot centered relationships, not dominion over nature, creating a site "that radically respatializes the body and notions of the self" (King 2019, 130). Within the backdrop of the plantation, their knowledge and seeds were utilized for plantation commodities and provisions on the plot. As Carney and Rosomoff (2009, 154) state, "Slavery signifies not only an appropriation of the body and its labor but also of the knowledge and ideas held by enslaved beings." However, through fugitivity, enslaved people utilized their ancestral tools and knowledge that formed into their folk culture as a tool of guerilla resistance against the ontology of the plantation. By retelling this history through the lens of radical Black ecology, we witness the birth of a new ecology rooted in African tradition that respatializes our sense of place and being in an otherwise disorienting, abusive system, and the succession of that ontology into heritage farming today.

Conclusion

In this chapter, we reorient the popular narratives of Black relationality to nature as one of violence, dispossession, oppression, and fungibility to the self-determined formations of our culture. We introduce "ontological succession" as a term defining how sociopolitical and ecocultural practices live on in our changing world and use it to not only trace the legacy of the plantation to the city but also to map the legacy of the plot to the lot. This process not only maps how the plantation became the city but, more importantly, how our community manifests our ancestral practices in our fugitive practices today.

Gazing at the fringes of these geographies, we guide you through the resistant legacy of our ancestors and their land-based practices. The ways they actively resisted the ontology of the plantation by engaging their ancestral ecocultural relationships in agriculture were passed down, informing following generations of our collective liberation practices. Today, we see this legacy live in the community agricultural movements of the garden lot in cities. By simply engaging with nature on our terms, particularly through our agricultural practices, Black growers are cultivating new ecologies of being. These formations are environmentally and socially just. This work is not new or novel. It is a part of the plot's logic of resistant, subversive work that recurs throughout Black history in this country. It is radical Black ecology in action.

References

Bartlett, K. 2022. "Reclaimed Land: Philadelphia's Urban Agriculture Movement." 34th Street. November 17. https://www.34st.com/article/2022/11/philadelphia-urban-agriculture-community-gardens-urban-tree-connection-norris-square-neighborhood-project-north-philly-peace-park.

Campt, T. 2014. "Black Feminist Futures and the Practice of Fugitivity." Barnard Center. https://bcrw.barnard.edu/videos/tina-campt-black-feminist-futures-and-the-practice-of-fugitivity/.

Carney, Judith, and Richard N. Rosomoff. 2009. *In the Shadow of Slavery: Africa's Botanical Legacy in the Atlantic World*. Berkeley: University of California Press.

City of Philadelphia Parks and Recreation. 2023. "Growing from the Root: Philadelphia's Urban Agriculture Plan Documents." City of Philadelphia. https://www.phila.gov/documents/philadelphias-urban-agriculture-plan.

Garth, Hannah, and Ashanté M. Reese. 2020. *Black Food Matters*. Minneapolis: University of Minnesota Press.

Gilbert, Charlene, and Quinn Eli. 2000. *Homecoming: The Story of African-American Farmers*. Boston: Beacon Press.

Gilmore, Ruth Wilson. 2007. *Golden Gulag*. Berkeley: University of California Press.

Hare, Nathan. 1970. "Black Ecology." *Black Scholar* 1 (6): 2–8. https://www.jstor.org/stable/41163443.

Harris, Jessica B. 2012. *High on the Hog*. New York: Bloomsbury.

Hartman, Saidiya, and Stephen Best. 2005. "Fugitive Justice." *Representations* 92 (1): 1–15. https://www.jstor.org/stable/10.1525/rep.2005.92.1.1.

Hartman, Saidiya. 2007. *Lose Your Mother: A journey along the Atlantic Slave Route*. New York: Farrar, Straus, and Giroux.

King, Tiffany L. 2019. *The Black Shoals: Offshore Formations of Black and Native Studies*. Durham, NC: Duke University Press.

McKittrick, Katherine. 2006. *Demonic Grounds: Black Women and the Cartographies of Struggle*. Minneapolis: University of Minnesota Press.

McKittrick, Katherine. 2013. "Plantation Futures." *Small Axe* 17 (3): 1–15. https://read.dukeupress.edu/small-axe/article-abstract/17/3%20(42)/1/33296/Plantation-Futures.

Opperman, Romy. 2020. "We Need Histories of Radical Black Ecology Now." AAIHS. https://www.aaihs.org/we-need-histories-of-radical-black-ecology-now/.

Sharpe, Christina. 2016. "The Wake." In *In the Wake: On Blackness and Being*, 1–24. Durham, NC: Duke University Press.

Simpson, Audra. 2016. "Consent's Revenge." *Cultural Anthropology* 31 (3): 326–333. https://doi.org/10.14506/ca31.3.02.

Story, K. A. 2015. Racing Sex—Sexing Race: The Invention of the Black Feminine Body. In *Imagining the Black Female Body: Reconciling Image in Print and Visual Culture*, edited by C. Henderson, 23–43. New York: Palgrave Macmillan US. https://doi.org/10.1057/9780230115477_2.

Strauss, Mariya. 2019. "Back to the Land: Struggles over Food and Real Estate Development in Philadelphia." *New Labor Forum* 28 (2): 79–83. https://doi.org/10.1177/1095796019836747.

Taylor, Dorceta E. 2014. *Toxic Communities*. New York: NYU Press.

Twitty, Michael. 2018. *The Cooking Gene*. New York: Amistad.

Washinton, Karen. n.d. "Frequently Asked Questions." Accessed April 2025. https://www.karenthefarmer.com/faq-index.

White, Monica M. 2018. *Freedom Farmers: Agricultural Resistance and the Black Freedom Movement*. Chapel Hill: University of North Carolina Press.

Wilkerson, Isabelle. 2010. *The Warmth of Other Suns*. New York: Penguin Random House.

Wynter, Sylvia. 1971. "Novel and History, Plot and Plantation." *Savacou* 5: 1–8.

5 Healing Colonial Wounds: Indigenous Food Sovereignty, Land Back, and Cultural Resurgence

Charlotte Coté (Tseshaht Nuu-chah-nulth)

My English name is Charlotte Coté; my quu'as or Indigenous name is łuutiisma?uƛ (Carrying Thunder). I am Tseshaht. We are part of the larger nation of Nuu-chah-nulth (located on the west coast of Vancouver Island in British Columbia, Canada). The name of our ancestral territory is ćuumaʕas and refers to the name of the river that runs through our community.[1]

For thousands of years before contact with Europeans, Indigenous people held autonomy over their food systems and maintained food security through a rich knowledge of their environment and food resources, passed down through oral tradition and long-standing land stewardship and cultivation practices (Turner and Turner 2008). Reviving and restoring our Indigenous foodways by enacting food sovereignty is directly linked to decolonization by resisting Western and unhealthy foods that were forced on us—forms of culinary imperialism and food hegemony. For my Tseshaht people, and Indigenous people worldwide, re-Indigenizing our diets is at the heart of the cultural resurgence and food justice movements we are witnessing in our communities today.

I begin this chapter with a story of one of my favorite harvesting traditions, picking wild trailing blackberries. We Indigenous peoples come from oral traditions, and our knowledge, theories, and epistemological frameworks are developed, maintained, and transferred from one generation to the next through the stories we tell and share. Our stories are both individual and communal and are grounded in our own personal lived experiences, while simultaneously, connecting us to our families, communities, and cultures in a deep and profound way. As Cree/Saulteaux scholar Margaret Kovach (2009)

---

quu?us ?imtii Charlotte Coté.?uuklahmaḥ łuutiisma?uƛ. ćiša?aqsumaḥ. nuučaañu uł?atḥin. histaqšiƛaḥ ćuumaʕas.

explains, "Stories remind us of who we are and of our belonging. Stories hold within them knowledges while simultaneously signifying relationships . . . [stories] tie us to our past and provide a basis for continuity with future generations." In other words, the stories we tell and the stories we hear help shape our identity, who we are—our stories are Indigenous theory "in action" (Million 2009).

Shut Up! We're Bonding!

On a hot summer day, my aunty Misbun[2] and I were out doing what we always did since I was a young girl—we were driving on winding, dusty gravel roads up steep mountains in the ḥaḥuułi, ancestral homelands, of my people the Tseshaht, on the west coast of Vancouver Island, on our quest for the sweet and delectable qaałqaawi, wild trailing blackberry (Kuhnlein and Turner 1991; Turner 2005).[3] These qaałqaawi are quite different than the tupkupeeh, Himalayan blackberry,[4] that were brought here from Europe and grow alongside the roads close to urban centers. The qaałqaawi is much smaller, hardier, and considerably richer in antioxidants than their tame cousins and are high in vitamin C, potassium, magnesium, and other flavonoid phytochemicals (Kuhnlein and Turner 1991). They grow high in the mountains in flat areas within the dense coastal rainforests that blanket the Northwest Coast (Turner 2005).

It almost seems like a misnomer to call qaałqaawi "wild." The Nuu-chah-nulth-aht, (Nuu-chah-nulth people), like other Indigenous peoples, cultivated lands in our ḥaḥuułi to encourage the growth of certain food plants, and this was done through sophisticated selective and controlled burning, promoting the growth and production of berry vines (Kuhnlein and Turner 1991; Turner 2005). But through colonization and forced placement on reserve lands, we lost many of our traditional harvesting sites, which fell under the control of forestry companies, and as a result, we lost our ability to engineer the lands for production. However, this ancestral knowledge continued to be transferred through the generations, and my grandfather Hughie Watts (aunty Misbun's father) was raised with the understanding and ecological knowledge of land burning. Grandpa would drive through our former harvesting sites and take note of where the forestry companies were slash burning to remove the underbrush to make it easier to cut trees and remove the fallen logs. He knew that in a few years these would be ideal growing areas for the qaałqaawi.

Grandpa passed on his cultural knowledge to us, especially the knowledge of where to find the best ńačyuu, or berry patch. But this afternoon was not the day for finding a good ńačyuu, and there we were hours later, still looking for one. The day shifted into the late afternoon and I was getting tired of looking for qaałqaawi, what seemed to be a hopeless situation, and I was about ready to tell my aunty we should just give up and go home. And then my aunty steered around a slight bend in the gravel road and there it was a few yards ahead of us on the right side of the road—a ńačyuu! My aunt slammed on the brakes and then hastily pulled over to the side of the narrow road. Since the sun was already going down on this side of the mountain, we needed to hurry. We grabbed our berry-picking pails out of the trunk of the car and ran to the small patch of ripe, deep purple qaałqaawi just waiting to be plucked from their hardy vines. I found a comfortable spot close to the road, and aunty Misbun made her way through the vines to start picking from the other side of the patch about 20 feet away. And then we began plucking berries to our hearts content. My head was down, my fingers were going a mile a minute, and my thoughts were lost in berry-picking heaven.

We had been picking for about fifteen minutes when I suddenly felt my aunt rush past me. I pulled my head up from the vines to ask my aunt what the heck was going on. "What's the matter?" I queried. Sometimes when we were picking berries, we would run into bears that were also enjoying a berry feast and, while they usually did not bother us, we had to be cautious if it was a mother with her cubs, who you did not want to come close to. But I looked around and did not see any. "Bees, bees!" aunty Misbun yelled, madly swatting at the air around her head as she ran toward our vehicle. My aunt had stepped right into the middle of a hornet's nest. "What the hell!" I gasped. I grabbed my pail and stumbled to my feet, spilling all the juicy berries I was dreaming about eating when I got home.

I sprinted behind my aunt while swatting at the bees that were now after me. We got to the vehicle and jumped inside. Aunty Misbun started the car and we drove off leaving behind the bees—and leaving behind most of our qaałqaawi. By this time, I had enough and was ready to call it a day. I was tired, I was hungry, and I was aching from the couple of bee stings I received. But not my aunty Misbun. She wanted her qaałqaawi, and the boiling sun, dust, dirt, hunger, and a few bee stings were not going to keep her from getting them. I looked at my aunt and said, "It's hopeless. Why don't we just go home? This is crazy. We're never going to find any more berries!" My aunt responded, "Shut up! We're bonding!" And, with a trickle of sweat running down her cheek, a slight smile on her lips, and a persistent look in her eyes, she continued driving up the mountain.

Settler Colonialism and Indigenous Health

Many years after I started my research on Indigenous foodways, I began to reflect on what my aunty Misbun meant when she said we were "bonding" during one of our berry-picking adventures. My aunty intuitively understood the cultural and familial importance of this tradition and how picking berries together not only reinforced our bond with each other as aunt and niece but also maintained and reinforced our bond to this important cultural food tradition, to our qaałqaawi (blackberry) relatives, to the nisma (land), and to our nonhuman kinfolk we encountered while on our berry-picking excursions.

It was not until I was an adult that I realized how fortunate I was to have been raised with our haʔumštup, cultural foods, and our food traditions, and that I was raised to understand the importance of staying connected to these foods, which I have done throughout my life. As an Indigenous food studies scholar born and raised in my Tseshaht community, I have studied, witnessed, and experienced how colonization and the perpetuation of settler colonialism through boarding schools, habitat destruction, socioeconomic marginalization, and the imposition of a Western diet have impacted my people's physical, nutritional, and spiritual health and caused us—like all Indigenous peoples globally—to be food insecure.

A 2017 study estimates that one in five deaths globally are associated with the Western diet, which is high in salt, sugar, and fat, as well as red and processed meats. According to the study, eleven million deaths were attributed to a poor diet, creating the highest risk factor in the world (GBD Diet Collaborators 2019). Worldwide, there has been an increase in disease in the global population. The industrialization and commodification of foods globally have a direct correlation with the rise of disease, which is more specifically linked to the increase in highly processed foods and refined grains and the use of chemicals to raise crops and animals in monoculture methods (Pollan 2008).

Indigenous people bear a disproportionate brunt of the burden of health and social suffering, greater than the general population, because of colonization and the omnipresent perpetuation of settler colonialism that disrupts our sacred relationships to our homelands and waters that provide us with nutritious cultural foods (Egeland and Harrison 2013). Settler colonialism is a political, legal, economic, and social structure that, as Glen Coulthard (Yellowknives Dene) explains, is structurally oriented around "the dispossession of Indigenous peoples from their lands" and "their self-determining authority" over their homelands (Coulthard 2014). When thinking about how settler colonialism operates, you can envision it as an "erasure" to "disappear Indigenous peoples," to erase our very existences (Tuck and Yang 2012). The erasure executed through federal Indian assimilation policies was designed to destroy our political, economic, social, and spiritual systems. The colonial erasure denied and ignored our histories and invented and perpetuated myths and stereotypes that create spaces where ignorance and racism can fester. As a result of social and economic inequality, and systemic racism, Indigenous peoples, as well as people of color, continue to struggle for racial, social, environmental, and food justice.

One of the most insidious policies designed to erase us was the federal Indian boarding school system, developed in the late 1800s in both the US and Canada, in which Indigenous children between the ages of five and fifteen were removed from their families and communities and placed in these segregated schools for ten to twelve months out of the year. In these schools, Indigenous children faced horrific physical, emotional, and sexual violence. Many scholars, including myself, have written about the appalling abuses that children suffered in these schools, and in the last twenty years the "survivors" of these schools have shared their stories of trauma in an attempt to heal the pain and reconcile with this horrific past (Adams 1995; Nuu-chah-nulth Tribal Council 1996; Reyner and Eder 2004; Million 2009). Trauma is deeply embedded in our collective Indigenous experience. As tribal societies, Indigenous identity to a large extent is a collective identity, so the experiences of the generations of children who attended these schools not only affected those students but also resulted in "collective historical trauma," what Dian Million (Tanana Athabaskan) (2013) so aptly describes as a colonial "wound" leading to the contemporary health issues we experience in our Indigenous communities today.

In my book, *A Drum in One Hand, A Sockeye in the Other* (2022), I write about a boarding school that was erected in my community, a ten-minute walk from where I grew up. While this book focuses on Northwest Coast Indigenous food sovereignty, I felt it was important for people to understand how the boarding school system dramatically impacted Indigenous health issues through forced dietary change on the Indigenous children attending these schools. In the schools, Indigenous children were required to eat foods that many had never eaten before, foods such as domesticated meats and processed and unhealthy foods, while at the same time estranging them from their own cultural and nutritious foods (Coté 2022; Truth and Reconciliation Commission of Canada 2015; CBS News 2021).

Federal Indian policies such as the boarding school system and settler colonialism have had profound impacts on Indigenous peoples' health and food security worldwide. The majority of Indigenous diets have been overwhelmed by industrial foods and a Western diet of foods high in salt, sugar, and fat, with traditional foods playing a supplemental role. This has contributed to the alarming rise of lifestyle diseases such as diabetes, heart disease, and autoimmune disease, which for many Indigenous communities has reached epidemic levels (Blue Bird Jernigan et al. 2010; National Collaborating Center for Indigenous Health and Statistics Canada 2013; Gracey and King 2009; King et al. 2009; Egeland and Harrison 2013).

In the larger US population, we also see a rise in lifestyle disease correlating directly with the rise in the consumption processed foods leading to high rates of obesity. For Native American adults and children, obesity is at a higher rate than all the other races combined in the US (Story et al. 1999).

A Holistic Approach to Indigenous Food Sovereignty

Indigenous peoples continue to struggle for basic human rights and to achieve health and wellness within our respective communities. These struggles align with the challenges tribes face when trying to assert and/or exercise their sovereignty over their communities, over their homelands, and over their health. This connects to food sovereignty, a concept that refers the right to healthy and culturally appropriate food produced through ecologically sound and sustainable methods and the ability to harvest on your traditional lands and in your traditional waterways (Declaration of Nyéléni 2007)—a basic human right.

Prior to colonization, traditional hunting, fishing, and gathering activities provided nutrient-rich diets to Indigenous communities. Depending on what was available in the geographic area, these diets were high in animal proteins, animal fats, and fat-soluble vitamins. They also contained lots of plant sources such as wild rice, tubers, chenopods, beans, seeds, maize, squash, berries, and leafy vegetables (Mailer and Hale 2013). Postcolonization, however, there has been a continual shift in Indigenous diets toward processed foods (National Collaborating Center for Aboriginal Health and Statistics Canada 2013, 3–4). Like everyone else, Indigenous people were also pulled into the savvy marketing of multi-billion-dollar capitalist corporations to get their industrialized, processed foods on our tables and into our diets (Moss 2013).

Indigenizing food sovereignty moves beyond a rights-based discourse and places emphasis on Indigenous responsibility, mutuality, kinship, and relationships with the natural world (Coté 2016). Rebuilding these relationships is positioned within a restorative framework that Potawatomi scholar Robin Kimmerer contends creates duties and responsibilities for both humans and nonhumans. "Just as all beings have a duty to me, I have a duty to them," Kimmerer writes. "If an animal gives its life to feed me, I am in turn bound to support its life. If I receive a stream's gift of pure water, then I am responsible for returning a gift in kind" (2013, 115).

Indigenous food sovereignty emphasizes the importance of tribes having the ability to exercise political, economic, and cultural sovereignty and having access to our lands and waters, where we harvested, fished, and hunted. Indigenous peoples feel a sense of frustration when our voices are continually ignored when it comes to environmental or food security issues that impact our very existence and survival. Māori scholar Linda Tuhiwai Smith writes, "It angers us when practices . . . are still employed to deny the validity of indigenous peoples' claims to existence, to land and territories, to the right of self-determination, to the survival of our languages and forms of cultural knowledge, to our natural resources and systems for living within our environments" (2012, 1). Indigenous peoples not only face food injustice but also experience environmental injustice, threatening the very ecosystems that sustain our cultures and communities.

Potawatomi scholar Kyle Pows Whyte asserts that food injustice can manifest as a "violation of food sovereignty" and the perpetuation of settler colonialism that works to destroy Indigenous foods and foodways and undermine tribes' "collective self-determination" and what Whyte defines

Figure 5.1
Coté with her Aunty Marilyn preparing salmon for the smokehouse. Photo by Charlotte Coté.

as "collective continuance" over our homelands (2017, 3). At its most basic, environmental injustice is "the disproportionate exposure of communities of color, including Indigenous communities, and the poor to pollution, and its concomitant effects on health and environment, as well as the unequal environmental protection and environmental quality provided through laws, regulations, governmental programs, enforcement, and policies" (Mantay 2002, 161). Here in the Northwest Coast, pollution, ocean acidification, water and land contamination, urban development, pipelines, recreational activities, fish farms, habitat destruction, and biodiversity loss all continually threaten our Indigenous foodways and the environment in which these foods thrive. Environmental degradation is a key factor in the decline of our foodways, along with climate change, all of which contribute to uncertainty and instability in our food systems (Turner, Plotkin, and Kuhnlein 2013).

The way environmental injustice impacts Indigenous peoples and communities can be seen as "slow violence." Slow Violence "occurs gradually and out of sight, a violence of delayed destruction . . . that is typically not viewed as violence at all" (Nixon 2013, 2). Environmental issues such as climate change, oil spills, deforestation, the thawing of icebergs, and ocean acidification are all having drastic impacts on Indigenous communities. "The disproportionate impacts of environmental contamination on Indigenous Peoples and communities of color are the basis of the now well-accepted concept 'environmental racism'" (Women's Earth Alliance and Native Youth Sexual Health Network 2016).

Indigenous peoples continually strive for environmental justice, which is defined as the "fair treatment and meaningful involvement of all people regardless of race, color, national origin, or income, with respect to the development, implementation, and enforcement of environmental laws, regulations, and policies" (US EPA 2023, 13). For Indigenous people like Swinomish tribal member Shelly Vendiola, who has been leading the environmental justice fight in the Pacific Northwest, environmental justice policies also need to include "valuing the health of our communities and the earth as well as protection from discrimination, dispossession, and exploitation" (personal communication, May 12, 2020).[5] Alison Alkon and Julian Agyeman highlight how "low-income people and people of color have been systematically denied access to the means of food production, and are often limited in their abilities to consume healthy foods" (2011, 331; see the introduction to this volume). For Indigenous people, food justice not only means having the ability to harvest and consume our cultural foods; it also means having access to our ancestral lands and waters where our foods thrive, which is at the core of the Land Back Movement (Tuck and Yang 2012).

If we look at decolonization as the undoing of settler colonialism, then decolonization needs to bring about the repatriation of Indigenous lands. Loss of homelands has created a situation where many tribal members cannot live in their communities because there is not enough land, this correlating with the rise of Indigenous urbanization, where more than 50 percent of Indigenous people in the US and Canada live in urban centers today. The Land Back Movement is about reclaiming Indigenous lands so that we reengage with these lands in a cultural and respectful way, especially in supporting our food sovereignty efforts. Canada Council for the Arts chair Jesse Wente contextualizes Land Back within the context of strengthening tribal sovereignty by returning decision-making over reclaimed lands to Indigenous peoples. Wente says, "It's about self-determination for our Peoples here that should include some access to the territories and resources in a more equitable fashion, and for us to have control over how that actually looks" (David Suzuki Foundation 2023).

Indigenous food sovereignty emphasizes the importance of having access to our lands and waters, where we harvested, fished, and hunted. The late activist Billy Frank Jr., a Nisqually tribal member and leader in the fight for Northwest Coast treaty fishing rights, positioned food sovereignty at the core of Indigenous struggles for political and cultural sovereignty. He said, "Our treaties recognize that food is at the center of our cultures. Indian tribes are sovereign nations, and part of that sovereignty includes access to the traditional foods needed to keep ourselves and our communities healthy and strong" (Frank 2012).

Land Back and reclaiming food sovereignty strengthen traditional ecological knowledge, the body of knowledge, belief, and practice that has been passed down through generations that honor the sacred relationships we have to our homelands and to our nonhuman kinfolk. For example, in the research I have conducted on Tseshaht foodways, I demonstrate how enacting food sovereignty is grounded in our Tseshaht/Nuu-chah-nulth principles of ʔiisaak, "being respectful," and ʔuʔaałuk, "to take care of." Our principle of ʔiisaak applies to all life-forms as well as to the land and water and, at its most basic understanding, teaches that all life-forms are held in equal esteem (Coté 2022; Turner et al. 2000).

Our relationships to the plants and animals that give themselves as food derives from ʔiisaak, which enforces sustainability and places sanctions on those who are stingy or wasteful (Turner et al. 2000). The underlying vision of ʔuʔaałuk is to "take care of" the ḥaḥuułi, ancestral homelands, in a way that is consistent with Nuu-chah-nulth values and principles, a responsibility given to our people through n̓aas, our Creator (Atleo 2004, 17–18, 130). The principles of ʔiisaak and ʔuʔaałuk are embedded within the overarching philosophy of ḥačatakma ćawaak (also written as hišukʔiš ćawaak) meaning "everything is interconnected," with the understanding that we must honor the wisdom and values of ancestral knowledge in maintaining responsible and respectful relationships with the environment. These guide the relationships we have to our natural and spiritual worlds.

This philosophy of ḥačatakma ćawaak connects people, animals, plants, and the natural and the supernatural or spiritual realms in a seamless and interconnected web of life where all life-forms are revered and worthy of mutual respect. The land, water, animals, and plants are regarded as your kinfolk, not as a commodity that can be exploited (Atleo 2004; Turner et al. 2000; Raymond et al. 2013). Embedded within these human and nonhuman relations is the understanding of responsibilities that Indigenous peoples strive to uphold in our social, cultural, and economic practices, responsibilities that have evolved into unwritten laws over millennia. Indigenous food sovereignty is encapsulated within an Indigenous philosophy and worldview that, unlike Western/Christian concepts of humans having authority over all living things, understands the human-ecosystem relationship as one of reciprocity and respect, where humans do not control nature but live in harmony with it. Indigenous food sovereignty therefore embodies a deep spiritual appreciation for food as a sacred gift (Morrison 2006).

For Indigenous peoples, reclaiming food sovereignty and reclaiming lands, or Land Back, means revitalizing the sacred relationships we have to our ancestral homelands, our waters, and our nonhuman kinfolk; being able to subsist on those lands and practice ceremonies that are sacred to those specific lands; and returning our communities to their ultimate holistic health.

Today, Indigenous communities throughout the US and Canada are experiencing a cultural resurgence and re-Indigenizing their diets through food sovereignty programs and initiatives to reconnect to our cultural foods and practices and to have access to our traditional territories to harvest them, which is fundamental to our health and our cultural survival. Enacting food sovereignty restores and strengthens individual and community health and wellness and is a way we can assert our cultural and political autonomy. Revitalizing our food practices has become a potent cultural symbol, as Indigenous peoples recognize that eating our cultural food and making and having the choice to eat these foods is a political act—a resistance to settler colonialism.

What non-Indigenous people might not immediately understand is that, for Indigenous people, our physical, spiritual, and emotional health are directly tied to our ability to eat our traditional foods. Restoring our traditional food practices by enacting food sovereignty allows us to experience a special connection to our cultures and our lands because every plant and animal carries their own spiritual gifts, so there is "a sense of vitality and belonging" that comes with eating the foods that provided our ancestors with optimum health and longevity (Krohn and Segrest, 9). Indigenous peoples have a cultural concept of "food as medicine," which promotes a holistic approach to maintaining and restoring the dietary, emotional, and spiritual health of our bodies. It also means maintaining the ecosystems that provide us with food (Morrison 2006).

In my research I explore healing in many ways, especially in relationship to our individual and collective physical, emotional, and spiritual health, what we Tseshaht define as tiičʕaqƛ, or holistic health. I also think through this in correlation to healing the *nisma̓* (land/earth) and our relationship to it. As Indigenous peoples, we live in a world of reciprocal relationships with the plants and animals that provide us with food, clothing, shelter, and cultural and spiritual sustenance, and we are bound by what Robin Kimmerer calls "a covenant of responsibility" where we treat plants, water, land, and air as gifts from the Creator (Kimmerer 2013, 115).

For Indigenous peoples, decolonization necessitates healing—healing ourselves, healing our communities, healing our traditional food systems, healing the lands, waters, and ecosystems and our relationships to them and our nonhuman kinfolk. Looking at the larger picture including both the food sovereignty and land reclamation movements—as well as the cultural revitalization, food justice, and decolonization movements we are experiencing in our Indigenous communities today—it is also about healing past wrongs. And I think about this with respect to allyship and collaboration with non-Indigenous peoples as we strive for self-determination and restoring healthy and thriving cultures, lands, and regenerative economies. If we can move together in a united way, as allies, then it is really about healing these colonial wounds that we Indigenous peoples have suffered and continue to suffer—to mend these wounds and work together in restoring healthy respectful and reciprocal relationships to our environment that will benefit us all as a global family.

Phonetic Key

a	sounds like the "a" in *what*
aa	sounds like a long, drawn out "a" as in *father*
c	sounds like the "ts" in *hats*
c̓	a glottalized sound, like "ts" but pronounced forcefully
č	sounds like the "ch" in *church*
č̓	a glottalized sound, like "ch" but pronounced forcefully
e	has the sound in *pet*
ee	has the sound in *eggs*
h	has the sound in *house*
ḥ	sounds like an "h" made deep in the throat like when breathing on glasses to clean them
i	sounds like the "i" in *it*
ii	sounds like a long "e," as in *greed*
k	sounds like the "k" in *kite*
kʷ	sounds like the "qu" in *quick*
k̓	a glottalized sound, like "k" but pronounced forcefully
k̓ʷ	a glottalized sound, like "qu" but pronounced forcefully
ł	the barred "l," a hissed version of "l," resembling the English "th" sound
ƛ	the barred lambda, sounds like "tl"
ƛ̓	a glottalized sound, like "tl" but pronounced forcefully
m	sounds like the "m" in *mother*
m̓	a glottalized sound, like "m" but pronounced forcefully
n	sounds like the "n" in *nose*
n̓	a glottalized sound, like "n" but pronounced forcefully
oo	has the sound in *paw*
p	sounds like "p" in *pig*
p̓	a glottalized sound, like "p" but pronounced forcefully
q	sounds like a "k" made deep in the throat
qʷ	sounds like a "k" made deep in the throat with a "w"
s	has the sound in *six*
š	sounds like the "sh" in *shoe*
t	sounds like the "t" in *toast*
t̓	a glottalized sound, like "t" but pronounced forcefully
u	has the sound in *took*

uu	has the sound in *boot*
w	sounds like the "w" in *wood*
w̓	a glottalized sound, like "w" but pronounced forcefully
x	the front "x," sounds like a cat's hiss
xʷ	like the front "x" but with a "w"
x̣	the back "x," a sound from the back of your throat, like clearing phlegm
x̣ʷ	like the back "x" but with a "w"
y	sounds like the "y" in *yes*
y̓	a glottalized sound, like "y" but pronounced forcefully
ʔ	the glottal stop, a closing of the vocal chords
ʕ	the pharyngeal stop, made by tensing the throat as if being strangled

Notes

1. Some of the material in this chapter has been adapted from my recent book, *A Drum in One Hand, a Sockeye in the Other: Stories of Indigenous Food Sovereignty from the Northwest Coast* (Seattle: University of Washington Press, 2022).

2. This is my aunt's nickname that she has had since she was a young girl. Her name is Eileen (Watts) Haggard. Her Tseshaht name is ḥeečis.

3. The trailing blackberry, known as *Rubus ursinus*, is found in open forests in western North America. The berries were harvested in July and either eaten immediately or dried for later use.

4. Nuu-chah-nulth spelling, *tuupkaappiiḥ*. The Himalayan blackberry, known as *Rubus discolor, Rubus armeniacus*, and *Rubus procerus*, is native to Asia and was brought to North America in the 1880s.

5. Michele (Shelly) Vendiola was a guest speaker in my Indigenous Resistance and Resurgence virtual class on May 12, 2020.

References

Adams, David W. 1995. *Education for Extinction: American Indians and the Boarding School Experience, 1875–1928*. Lawrence: University of Kansas Press.

Alkon, Alison Hope, and Julian Agyeman. 2011. *Cultivating Food Justice: Race, Class, and Sustainability*. Cambridge, MA: MIT Press.

Atleo, E. Richard, Umeek. 2004. *Tsawalk. A Nuu-chah-nulth Worldview*. Vancouver, Canada: UBC Press.

Blue Bird Jernigan, Valerie, Bonnie Duran, David Ahn, and Marilyn Winkleby. 2010. "Changing Patterns in Health Behaviors and Risk Factors Related to Cardiovascular Disease Among Indians and Alaska Natives." *American Journal of Public Health 100* (4): 677–683.

CBS News. 2021. "Your Questions Answered About Canada's Residential School System." https://www.cbc.ca/news/canada/canada-residential-schools-kamloops-faq-1.6051632.

Coté, Charlotte. 2016. "Indigenizing' Food Sovereignty. Revitalizing Indigenous Food Practices and Ecological Knowledges in Canada and the United States." *Humanities 5* (3): 1–14. https://www.mdpi.com/2076-0787/5/3/57/htm.

Coté, Charlotte. 2022. *A Drum in One Hand, A Sockeye in the Other. Stories of Indigenous Food Sovereignty from the Northwest Coast*. Seattle: University of Washington Press.

Coulthard, Glen, 2014. *Red Skin, White Masks. Rejecting the Colonial Politics of Recognition*, Minneapolis: University of Minnesota Press.

David Suzuki Foundation. 2023. *What is Land Back?* https://davidsuzuki.org/what-you-can-do/what-is-land-back/#:~:text=Land%20Back%20is%20an%20Indigenous,about%20the%20decision%2Dmaking%20power.

Declaration of Nyéléni, Declaration of the Forum for Food Sovereignty, Sélingué, Mali, February 27, 2007. https://nyeleni.org/IMG/pdf/DeclNyeleni-en.pdf.

Egeland, Grace M., and Gail G. Harrison. 2013. "Health Disparities: Promoting Indigenous Peoples' Health Through Traditional Food Systems and Self-determination." In *Indigenous Peoples' Food Systems & Well-being. Interventions & Policies for Healthy Communities*, edited by V. Harriet Kuhnlein, Bill Erasmus, Dina Spigelski, and Barbara Burlingame, 9–22. Rome: Food and Agriculture Organization of the United Nations.

Frank, Billy Jr. "Traditional Foods Are Treaty Foods." Northwest Treaty Tribes. March 5, 2012. https://nwtreatytribes.org/traditional-foods-are-treaty-foods/.

GBD Diet Collaborators. 2019. "Health Effects of Dietary Risks in 195 Countries, 1990–2017: A Systematic Analysis for the Global Burden of Disease Study 2017." *Lancet* 393: 1958–1972.

Gracey, Michael, and Malcolm King. 2009. "Indigenous Health. Part 1: Determinants and Disease Patterns." *Lancet, 374*, no. 9683: 65–75.

Kimmerer, Robin Wall. 2013. *Braiding Sweetgrass. Indigenous Wisdom, Scientific Knowledge, and the Teachings of Plants*. Minneapolis, MN: Milkweed Editions.

King, Malcolm, Alexandra Smith, and Michael Gracey. 2009. "Indigenous Health. Part 2: The Underlying Causes of the Health Gap." *Lancet* 374 (9683): 76–86.

Kovach, Margaret. 2009. *Indigenous Methodologies. Characteristics, Conversation, and Context*. Toronto: University of Toronto Press.

Krohn, Elise, and Valerie Segrest. 2010. *Feeding the People, Feeding the Spirit. Revitalizing Northwest Coastal Indian Food Culture.* Centralia, WA: Gorham Printing.

Kuhnlein, Harriett V., and Nancy J. Turner. 1991. "Traditional Plant Foods of Canadian Indigenous Peoples. Nutrition, Botany, and Use." Amsterdam: Oversees. https://openknowledge.fao.org/server/api/core/bitstreams/02134cf4-156b-47c7-972d-cf2690df1b55/content.

Mailer, Gideon, and Nicola Hale. 2013. "Decolonizing the Diet: Synthesizing Native-American History, Immunology, and Nutritional Science." *Journal of Evolution and Health* 1 (1): 1–41. https://escholarship.org/content/qt83n957nb/qt83n957nb.pdf?t=q3sazi.

Mantay, Juliana. 2002. "Mapping Environmental Injustices: Pitfalls and Potential of Geographic Information Systems in Assessing Environmental Health and Equity." *Environmental Health Perspectives* 110 (Suppl 2): 161–171.

Million, Dian. 2009. "Felt Theory: An Indigenous Feminist Approach to Affect and History." *Wicazo Sa Review* 24 (2): 53–76.

Morrison, Dawn. 2006. Final Report. 1st Annual Interior of B.C. Indigenous Food Sovereignty Conference. Final Report. https://www.indigenousfoodsystems.org/sites/default/files/resources/IFSC2006FinalReport.pdf.

Moss, Michael. 2013. *Salt, Sugar, Fat. How the Food Giants Hooked Us.* New York: Random House.

National Collaborating Center for Aboriginal Health, and Statistics Canada. 2013. "An Overview of Aboriginal Health in Canada." Prince George, Canada: University of British Columbia.

Nixon, Rob. 2011. *Slow Violence and the Environmentalism of the Poor.* Cambridge, MA: Harvard University Press.

Nuu-chah-nulth Tribal Council. 1996. *Indian Residential Schools: The Nuu-chah-nulth Experience.* Port Alberni, Canada: Nuu-chah-nulth Tribal Council Indian Residential School Study, 1992–1994.

Pollan, Michael. 2008. *In Defense of Food: An Eater's Manifesto.* New York: Penguin Books.

Raymond, Christopher M., Gerald G. Singh, Karina Benessaiah, Joanna R. Bernhardt, Jordan Levine, Harry Nelson, Nancy J. Turner, Bryan Norton, Jordan Tam, and Kai M. A. Chan. 2013. "Ecosystem Services and Beyond: Using Multiple Metaphors to Understand Human-Environment Relationships." *BioScience* 63 (7): 536–546.

Reyner, Jon, and Jeanne Eder. 2004. *American Indian Education: A History.* Norman: University of Oklahoma Press.

Smith, Linda Tuhiwai. 2012. *Decolonizing Methodologies. Research and Indigenous Peoples*. 2nd ed. London: Zed Books.

Story, Mary, Marguerite Evans, Richard R. Fabsitz, Theresa E. Clay, Bonnie Holy Rock, and Brenda Broussard. 1999. "The Epidemic of Obesity in American Indian Communities and the Need for Childhood Obesity-Prevention Programs." *American Journal of Clinical Nutrition* 69 (4): 747–754.

Truth and Reconciliation Commission of Canada. 2015. *Honouring the Truth, Reconciling for the Future. Summary of the Final Report of the Truth and Reconciliation Commission of Canada*. Truth and Reconciliation Commission of Canada. http://publications.gc.ca/site/eng/9.800288/publication.html.

Tuck, Eve, and K. Wayne Yang, 2012. "Decolonization Is Not a Metaphor." *Decolonization, Indigeneity, Education & Society* 1: 1–40.

Turner, Nancy J. 2005. *The Earth's Blanket. Traditional Teachings for Sustainable Living*. Seattle: University of Washington Press.

Turner, Nancy, Marianne Boelscher Ignace, and Ronald Ignace. 2000. "Traditional Ecological Knowledge and Wisdom of Aboriginal Peoples in British Columbia." *Ecological Applications* 10 (5): 1275–1287.

Turner, Nancy J., Mark Plotkin, and Harriet V. Kuhnlein. 2013. "Global Environmental Challenges to the Integrity of Indigenous Peoples' Food Systems." In *Indigenous Peoples' Food Systems & Well-being. Interventions & Policies for Healthy Communities*, edited by V. Harriet Kuhnlein, Bill Erasmus, Dina Spigelski, and Barbara Burlingame, 23–38. Rome: Food and Agriculture Organization of the United Nations.

Turner, Nancy J., and Katherine L. Turner. 2008. "'Where Our Women Used to Get the Food': Cumulative Effects and Loss of Ethnobotanical Knowledge and Practice; Case Study from Coastal British Columbia." *Botany* 86: 103–115.

US Environmental Protection Agency (EPA). 2023. Environmental Justice. https://www.congress.gov/crs-product/R47920#.

Whyte, Kyle Pows. 2017. "Food Sovereignty, Justice and Indigenous Peoples: An Essay on Settler Colonialism and Collective Continuance." In *Oxford Handbook on Food Ethics*, edited by A. Barnhill, T. Doggett, and A. Egan, 1–21. Oxford: Oxford University Press.

Women's Earth Alliance and Native Youth Sexual Health Network. 2016. "Violence on our Lands, Violence on our Bodies. Building an Indigenous Response to Environmental Violence." http://landbodydefense.org/uploads/files/VLVBReportToolkit2016.pdf. Accessed 12/4/2023.

II Confrontations

6 From Food Deserts to Food Apartheid: Origins, Histories, and Solutions

Gabriel R. Valle and Greig Tor Guthey

In the middle of the Obama administration's first term, First Lady Michelle Obama launched the "Let's Move" campaign to improve food access and combat childhood obesity, noting in a speech that 23.5 million Americans lived in food deserts (Obama White House 2010). "Our goal," Mrs. Obama said in her 2010 speech, "is to eliminate food deserts in America completely in seven years." With such an ambitious statement, food justice seemed to be on the horizon in the US. Yet, in 2022, Harvard University's Social Policy Lab found that 23.5 million Americans still lived in food deserts and 54 million Americans had no reliable access to food (Ney 2022). Even today, the "Let's Move" campaign's goal has yet to be reached fourteen years after Michelle Obama set out to solve the problem. It is not for wont of trying; Mrs. Obama continues to work toward healthier food for children; activists continue their work in food pantries, farms, and community gardens across the country; and students sit in our classrooms each semester hoping to learn how to address food insecurity. Access to healthy food is a much more complicated issue to solve than it was first assumed to be by everyone involved—Mrs. Obama, USDA officials, and many of us relatively privileged researchers.

Perhaps the proposed solutions for improved food access were not equipped to deal with the historical origins, depth, and complexities of the problem. For example, National Black Food and Justice Alliance founder Dara Cooper argues the food desert lens is a "dominant narrative that perpetuates [a] narrow analysis" (Cooper 2019, xi), which neglects both the historical formation of urban areas that resulted in the lack of access to quality food and how communities mobilize the assets they do have in response. While there is an enduring literature on food deserts, especially in the fields of policy and planning (Zurawski 2023), there are new efforts to reframe and alter the analysis. Prominent food justice activist Karen Washington denounced the idea of food deserts and began to call these circumstances "food apartheid" (Brones 2018), a term that explicitly confronts past and present white supremacies and the ways they are/were inscribed on the landscape. We follow her lead in using this term to highlight how food insecurity is not naturally occurring but results from the historical consequences of segregated communities and ongoing uneven economic development (see chapter 7). In other words, food apartheid is the current structure of food exclusion in the US based on multiple factors such as race, class, age, gender, geography, and so on. Food apartheid is about more than accessibility; it is about how social structures systematically deny opportunity to some while providing it to others (see chapter 4). In this chapter, we suggest that the solution to the food inequality question may have less to do with *where* people acquire food and more with *how* the problem of food access is framed in the first place. We use the lens of food apartheid to reveal several processes of uneven, racialized, capitalist, and colonial development through which unequal food access was created and is sustained. To effectively confront food exclusion in the US and develop more lasting solutions, we suggest that a food apartheid lens provides a more comprehensive understanding of the multiple social causes of food exclusion. However, our list of causes is not comprehensive, so we invite you to consider other possible forms of exclusion that perpetuate food apartheid in communities across the country (see chapter 20).

Food Deserts

For over three decades, researchers and activists have worked on food deserts. Research on food deserts originated in Scotland in the 1990s to describe how poor access to affordable good-quality food resulted in unhealthy diets (Cummins and Macintyre 2002). The term grew in popularity in the early 2000s, and in the 2008 Farm Bill, the US Congress directed the US Department of Agriculture (USDA) to assess the extent of areas of the country with limited access to a variety of healthy and nutritious food (USDA 2012, 1). To make this assessment, USDA researchers looked for areas of the country in which people lived either more than one mile from a grocery store in urban contexts or more than 10 miles away in rural contexts. This became the working definition of a food desert. Later work by the USDA refined this understanding by linking it with income such that, for the USDA, the definition of a food desert became "low-income census tracts with a substantial number or share of residents with low levels of access to retail outlets selling healthy and affordable foods" (Ver Ploeg et al. 2011, paragraph 5).

By identifying areas as food deserts, federal, state, and local funds can be used to bring food into the city through grocery stores and various other markets and food initiatives. Funds can help establish community gardens, farm-to-school programs, and many other well-intended programs. There is even a tool—the USDA Food Access Research Atlas—to help identify food deserts by considering the distance one must travel to acquire good-quality food. Yet, as the Let's Move campaign helps illustrate, perhaps the barriers to good-quality food are not just the borders drawn around a low-income area with no grocery stores.

For example, the Deanwood community in Washington, DC, is a prototypical food desert. The community consists of brick and wood-framed houses with large front yards. It is a predominantly low-income community comprising mostly African American and other people of color. According to the USDA's Food Access Research Atlas, many who live there lack access to fresh and nutritious foods. Ashanté Reese (2019), who conducted extensive research in the region, describes the neighborhood as showing clear signs of urban decline and disinvestment, which are fundamental elements of a food desert. The term "food desert" almost automatically invokes images of remote places devoid of meaning.

Yet a community, just like a desert, is never truly empty, as Reese shows. They are always full of life, connection, and creation in its various forms. Beyond the "nothingness" of the food desert lens lies a more complex story of how and why food access has become so challenging for so many, but we need the right framing to see it. The focus on travel distance and income does not provide a sense of the fabric, vitality, and challenges neighborhoods like Deanwood face, nor does it reflect how residents respond to the material reality of not having access to good-quality food. When not critically examined, the food desert approach's emphasis on lack is a racialized framing that is "nearly synonymous with Blackness in much social science research" (Reese 2018, 412). Regardless of the physical distance one must travel to acquire food, Reese points out that people living in Deanwood are developing their own strategies and solutions to the barriers they face.

The food desert approach assumes everyone in a community experiences the hardship of distance to a grocery store in the same manner. But for people living in Deanwood, their own strategies mitigate many of the problems they face. Indeed, as Kolb (2021), Reese (2019), Alkon et al. (2013), and others have shown, choices about where to shop are not made based on proximity alone, as the USDA Food Access Atlas assumes. In fact, Kolb insists that food access inequality is one manifestation of a larger community investment problem he calls "retail inequality," which is the "result of purportedly 'color-blind' practices that stripped urban Black neighborhoods of the collective wealth necessary to support 'good retail'" (2021, 8–9). Reflecting on his years of research, Kolb insists that he came to realize that "the food desert fight was about something bigger than food" (2021, 177). It is about demanding retail that supports the needs and desires of a community rather than the interests of outside investors. The food desert framework depicts communities with unequal access to good-quality food as "bad" and the people who reside in these communities as incapable of fixing them. The purportedly "color-blind" practices Kolb is referring to reinforce the "ways policy and practice attempt to erase poor people's desires and choices" (Reese 2019, 59) to justify outside intervention.

Another problematic assumption is that food deserts are natural outgrowths of the free market, which is at best an oversimplification. Food desert research shows that supermarket chains make decisions about the location of stores based on markets and income, but people access food in myriad ways revealing how people have agency and work to create their own spaces despite the challenges and limitations of place. Moreover, food deserts, like economies, are not natural phenomena: they are historically, socially, and spatially produced. Indeed, the lack of grocery stores in a community is not the problem, but is a symptom of a larger structural condition brought on by "racially exclusionary practices" (Bradly and Galt 2014).

So, how can we get a handle on the lingering reality of food access inequality in America today? If the food desert lens has not led to a solution or even an effective framework for analysis, then perhaps we need to reframe the problem as Cooper, Washington, and many others do. Rather than reducing complex problems to simple solutions, it is time to use food apartheid as a human-centered lens to understand the histories, systems, and institutions that influence the conditions people confront and struggle with everyday.

The Origins of Food Apartheid

Settler colonialism set the stage for food apartheid in the US because it established the conditions for racial identity and property to become deeply interrelated concepts (Harris 1993). Colonialism is often considered a process of establishing control over a given territory for the purpose of taxation or resource extraction, but settler colonialism "is a structure not an event" (Wolfe 2006), and its patterns of exclusion, violence, and erasure are embedded into how society is organized. Settler colonialism is "different from other forms of colonialism in that settlers come with the intention of making a new home on the land, a homemaking that insists on settler sovereignty over all things in their new domain" (Tuck and Yang 2012, 5). In this way, settler colonialism is a contest over a way of life "because the settlers make Indigenous land their new home and source of capital, and also because the disruption of Indigenous relationships to land represents a profound epistemic, ontological, cosmological violence" (Tuck and Yang 2012, 5).

In the US, settlers used land claims and the Discovery Doctrine to justify their settlement on so-called empty lands. They forced Indigenous peoples to relocate to remote reservations where they were expected to adapt to sedentary lifestyles and transform their diets to accommodate Euro-American foodstuffs. Later, federal legislation empowered the government to expand the boundaries of these enclosures through further settlement. Signed in 1862, the Homestead Act provided any adult US citizen, which excluded freed Black slaves until after the Civil War and the passing of the Fourteenth Amendment in 1866, the ability to claim up to 160 acres of surveyed government land in the western US. The only requirement was to "improve" the land by building a dwelling and cultivating the land. On the one hand, the Homestead Act provided an opportunity for the westward expansion of US economic interest. Along with settlers came farms, towns, cities, railroads, and markets. On the other hand, settlement erased centuries of Indigenous land management practices and communities. The intent of the Homestead Act was to provide opportunities for white Euro-American settlers to live off the land and to eradicate Indigenous peoples and their ways of living. African Americans were excluded from these land grants, along with markets, loans, technology, and new knowledge, all while facing the daily threats of lynching, segregation, and other forms of racial discrimination for the next hundred years. Even in the 1960s, at the time when the US Congress was passing Civil Rights bills, the USDA was systemically denying Black farmers access to loans, information, training, and leadership positions, a practice that had being going on for decades (Daniel 2013).

In what became known as the Great Migration, roughly six million Black Americans moved from former Confederate states in the rural South to cities in southern, midwestern, and western states from the 1910s to 1970s. This migration is often understood as a voluntary migration where Blacks sought freedom and economic opportunity in the North, yet this migration was also a form of forced displacement in response to the various forms of violence, exclusion, and deeply held racism in the Jim Crow South (Wilkerson 2010). Thus, the Great Migration displaced millions of Blacks from the land. Many ended up in segregated Black ghettos in the urban North in what would become food deserts, but the systemic nature of this displacement and exclusion suggests food apartheid, which like settler colonialism, is a structure woven into the very fabric of this county and embedded across the entire US. Food access inequality is not the result of market forces or an outcome of individual choice but the product of a complex system of exclusion.

Racial Capitalism and Food Apartheid in Urban America

The food desert lens conceptualizes the placement of grocery options as a result of social class. The logic of supply and demand assumes that stores go where the money is. Yet, when we switch lenses, food apartheid pushes us to explore why poverty is so racialized to begin with. Moving beyond the one-dimensional framework of the food desert approach reveals how markets, and opportunities in those markets, are highly racialized. Cedric Robinson (2000) refers to the overlay between markets, opportunities, and race as "racial capitalism." Robinson argues that "the development, organization, and expansion of capitalist society pursued essentially racial directions," and these racial directions "would inevitably permeate the social structures emergent from capitalism" (2000, 2). Food apartheid is one of the results of those "racial directions," and it has many historical antecedents.

Perhaps the most obvious historical example is the actual system of apartheid from which the concept of food apartheid is derived. The system of institutional racial segregation and discrimination associated with South Africa and Namibia was a specific political project to marginalize and segregate people of color that was implemented between 1948 and 1994. The results of this political project were stark; they include deep impoverishment and disempowerment of Black South Africans and limited resources devoted to housing and public infrastructure in Soweto and other Black townships. Apartheid was dismantled through decades of riots, imprisonment (chapter 20), killings, protest, and negotiation, but its legacies persist to this day.

Similarly, the US has its own multilevel system regulating race. While its impact is less well understood or even recognized today, its legacies are also long-standing. The next section highlights several key mechanisms and tools that prolong and deepen the layers of exclusion that form the framings of food apartheid.

Instituting Food Apartheid

In the mid-twentieth century, sociologist Gunnar Myrdal recognized segregation as a causal factor contributing to Black Americans' isolation, economic status, and social problems, but research and policy analysis of segregation fell out of fashion in the 1970s and 1980s. Douglas Massey and Nancy Denton (1993) explain that "few appreciate the depth of Black segregation or the degree to which it is maintained by ongoing institutional arrangements and contemporary individual actions [at the end of the twentieth century]." Most observers in the 1990s understood segregation as part of an already-rectified racist past due to 1960s-era civil rights legislation and the passage of time, yet our communities are more segregated today than they were thirty years ago (Menendian et al. 2021). By the end of the millennium, most saw segregation as the individual choice of African Americans and other people of color rather than as the result of any institutional or racist policy agenda. Nevertheless, several ongoing practices prolong segregation across the country and thus reinforce food apartheid today.

Redlining

A key mechanism of segregation mentioned in the food literature is redlining. Prior to the twentieth century, Black and white Americans lived in somewhat more integrated neighborhoods. "Studies of Black residential life in northern cities around the time of the Civil War reveal little systemic exclusion from white neighborhoods on the basis of skin color" (Massey and Denton 1993, 20). Indices of racial segregation point to *integration* across many northern cities (although they also show increasing rates of segregation during the same period), while southern cities had even less segregation despite Jim Crow laws at the time. The racial divides apparent in today's urban landscape, with some neighborhoods synonymous with Black ghettos, are an invention of twentieth-century segregationists. Post-WWII-era policies, judicial and development decisions, mortgage industry practices, and highway construction were at the forefront of the creation of countless ghettos and barrios across the country that suffer from food apartheid.

The term "redlining" emerged in the 1960s based on maps developed in the 1930s by the Home Owners' Loan Corporation (HOLC), which was created by New Deal legislation intended to address the social and economic dislocations of the Great Depression. In the twentieth century, only whites had access to federally backed mortgages; HOLC provided the spatial justification by creating maps of borrowing risk in major US cities and towns. The maps categorized neighborhoods into four categories of risk labeled A–D, with A designated "best" and D considered "hazardous." Each neighborhood was outlined with an appropriate color: green, blue, yellow, or red. All D-rated neighborhoods were outlined in red and were largely occupied by minorities, and so the term "redlining" became associated with low-income and minority neighborhoods. Redlining is the systematic denial or selective raising of prices on numerous financial services, including mortgages and loans, which mostly affected African Americans.

While redlining is an illegal practice today, its policies have "never been remedied and their effects endure" (Rothstein 2017, xvii). Many once-redlined communities currently experience higher levels of pollution, traffic, reported crime, social exclusion, and less access to fresh, organic, and culturally appropriate food. These same communities are also frequently still communities of color. Long-term persistent inequalities stemming from the nineteenth century help to explain the intergenerational effects that racially motivated mortgage and lending institutions can have on the health and well-being of communities of color in the twenty-first century. So, while apartheid is not often associated with US cities, using food apartheid as a lens of analysis draws attention to how past and present racially discriminatory political structures, policies, and practices influence food access and control across the US. As explained by Karen Washington, "food apartheid looks at the whole food system, along with race, geography, faith, and economics" to understand the effects food access inequality has on communities and the policy choices that might best address food inaccessibility in rural and urban communities across the country.

Restrictive Community Covenants

Another tool of segregationists that was more pervasive than redlining in some parts of the country was the restrictive covenant. Restrictive covenants are often used to set limits or standards on things like minimum square footage, allowable building styles, paint color, and land use, but they also were used to exclude Blacks and other people of color. These restrictions became known as racial covenants and identified who could and who could not live in certain neighborhoods. Often clauses prohibited property sales to people of color or limited sales only to whites. Such restrictions were initially attached to individual deeds but were later applied to entire housing developments so that neighbors could enforce them if necessary; developers' access to financing was sometimes contingent on implementing such covenants. So pervasive were restrictive covenants that by the 1960s Blacks were excluded from 98 percent of new homes and 95 percent of neighborhoods across the US (Slater 2021).

While restrictive covenants existed everywhere in the country, they were most prominent in California. Racial covenants limited the ability of minorities to purchase homes in neighborhoods with good schools, access to green spaces, and clean industry. More than one hundred lawsuits in Los Angeles between 1937 and 1948 attempted to enforce such restrictions on Black homeowners who had purchased homes in violation of these covenants (Rothstein 2017). The practice was found to be unconstitutional in the 1950s and was formally banned in the Civil Rights Act of 1968 (Title VIII of this act is known the Fair Housing Act of 1968), yet the practices continue today and reflect a "fundamental realtor belief . . . [that] a neighborhood would only be successful in the long run if it was socially homogenous" (Slater 2021, 63). Racial covenants are one set of legal practices used across the country to create spaces of ample food on the one hand and food exclusion on the other.

Redevelopment and Highway Construction

Racially planned redevelopment also contributed to the current conditions of food apartheid we see today. Many cities saw tremendous growth during the post-WWII years and targeted minority communities for redevelopment starting in the 1950s. Sometimes based on prior beliefs that stemmed from racially motivated understandings of minority communities, redevelopment agencies leveraged local, state, and federal funds to uproot low-income communities and communities of color. One example of this is Sacramento's West End, which city officials considered a blighted section of the city ripe for redevelopment. Beneath the racialized gaze of redevelopment, the community was a vibrant and important multiracial community and business district with deep ties among the residents who lived there. As redevelopment brought in outside money and influence, it weakened community ties as people moved to other portions of the city (Hernandez 2009).

As highway construction became a national security goal during the Cold War, minority communities were greatly affected, and the story of Sacramento's West End was repeated as most major highways came to border, bifurcate, or separate minority communities across the country. These communities were already undervalued and disempowered from decades of exclusionary practices, which made them more appealing to developers and investors. Yet, these so-called blighted neighborhoods did not just happen; they were justified by the federal HOLC program and implemented by local, state, and federal disinvestment (Kolb 2021). Redevelopment projects promote local investment and simultaneously displace others by pricing them out or simply taking the land through eminent domain. From Seneca Village in New York (the city's first Black community, on land that is now part of Central Park) to West Oakland (Fairfax et al. 2012) and from Saint Paul's Rondo community (Urban Land Institute 2018) to Atlanta's widening of Interstate 75 (Stone 1989), minority communities across the country fell victim to state-sponsored displacement and its inevitable outcome: food apartheid. Food justice scholarship has often documented these histories in order to contextualize contemporary efforts to increase food access (McClintock 2011; Alkon 2012; Myers 2022; Kolavalli 2023). While redlining was a process of creating artificial, or perhaps ideological lines on paper, redevelopment and highway construction projects proved to be valuable tools used in urban planning and public policy to create a built environment to separate people from each other and from opportunities, including "healthy" food.

Conclusion

In this chapter we have outlined how the food apartheid lens helps us better understand how different people confront the systemic denial of access to food by providing several ways that food access inequality is sustained in the US. Like the work of Agyeman and McEntee, who offer the lens of urban political ecology to "shed light on the symptoms of unjust access to food" (2014, 211), we encourage food justice activists, policymakers, researchers, and students to frame the issues associated with the unequal access to quality food through the lens of food apartheid to better recognize its historical origins and the many faces of oppression and to better grasp how communities express their agency.

Framing food access inequality as food apartheid encourages an entirely different analysis. Where the food desert approach might offer seemingly easy solutions, such as building grocery stores in low-income communities, food apartheid considers food access from a systemic and structural perspective where there are no easy solutions. "Food apartheid [is] a human-created system of segregation that relegates certain groups to food opulence and prevents others from accessing life-giving nourishment," writes Leah Penniman (2018, 4). Understanding how these divisions came to be points to a more extensive historical and geographic set of problems. Both terms—food desert and food apartheid—address the problem of food access, but the latter illuminates the deeper historical issues involved in US food access inequality and may lead to more comprehensive and lasting solutions. As Washington explains, "You say 'food apartheid' and you get to the root cause of some of the problems around the food system" (Brones 2018, para 12).

The origins of food access inequality are complex, and the framing provided by food apartheid enables a more comprehensive understanding of its nuances. When we reframe food access inequality as something more than market failure and beyond market-based fixes and assumptions about geographical proximity, we arrive at a place that allows us to see and understand the many different systems influencing the conditions of daily life. Food apartheid provides us a lens to understand the many faces of racial exclusion and settler colonialism that effect the structural conditions of food inequality and to see how people assert agency despite larger structures of power. Food deserts are not empty but full of community, history, and local solutions to the ongoing exclusionary practices of food apartheid.

References

Agyeman, Julian, and Jesse McEntee. 2014. "Moving the Field of Food Justice Forward Through the Lens of Urban Political Ecology." *Geography Compass*. 8(3): 211–220.

Alkon, Alison Hope. 2012. *Black, White, and Green: Farmers Markets, Race, and the Green Economy*. Athens: University of Georgia Press.

Alkon, Alison Hope, Daniel Block, Kelly Moore, Catherine Gillis, Nicole DiNuccio, and Noel Chavez. 2013. "Foodways of the Urban Poor." *Geoforum* 48 (August): 126–135.

Bradley, Katherine, and Ryan Galt. 2014. "Practicing Food Justice at Dig Deep Farms & Produce, East Bay Area, California: Self-Determination as a Guiding Value and Intersections with Foodie Logics." *Local Environment* 19(2): 172–186.

Brones, Anna. 2018. "Food Apartheid: The Root of the Problem with America's Groceries." *Guardian*, May 15, 2018. https://www.theguardian.com/society/2018/may/15/food-apartheid-food-deserts-racism-inequality-america-karen-washington-interview.

Cooper, Dara. 2019. Foreword to *Black Food Geographies: Race, Self-reliance, and Food Access in Washington, DC*, xi-xiii. Chapel Hill: University of North Carolina Press.

Cummins, Steven and Sally Macintyre. 2002. "Food Deserts"—Evidence and Assumption in Health Policy Making." *BMJ* 325(7361): 436–438.

Daniel, Pete. 2013. *Dispossession: Discrimination Against African American Farmers in the Age of Civil Rights*. Chapel Hill: University of North Carolina Press Books.

Fairfax, Sally K., Greig Tor Guthey, Louise Nelson Dyble, Lauren Gwin, and Monica Moore. 2012. *California Cuisine and Just Food*. Cambridge, MA: MIT Press.

Harris, Cheryl I. 1993. "Whiteness as Property." *Harvard Law Review* 106(8): 1707–1791.

Hernandez, Jesus. 2009. "Redlining Revisited: Mortgage Lending Patterns in Sacramento 1930–2004." *International Journal of Urban and Regional Research* 33 (2): 291–313.

Kolavalli, Chaaya. 2023. *Well-Intentioned Whiteness*. Athens: University of Georgia Press.

Kolb, Kenneth H. 2021. *Retail Inequality: Reframing the Food Desert Debate*. Berkeley: University of California Press.

Massey, Douglas S., and Nancy A. Denton. 1993. *American Apartheid: Segregation and the Making of the Underclass*. Cambridge, MA: Harvard University Press.

McClintock, Nathan. 2011. "From Industrial Garden to Food Desert." In *Cultivating Food Justice*, edited by Alison Alkon and Julian Agyeman, 89–120. Cambridge, MA: MIT Press.

Menendian, Ephen, Samir Gambhir, and Arthur Gailes. 2021. "The Roots of Structural Racism Project: Twenty-First Century Racial Residential Segregation in the United States." Othering and Belonging Institute. https://belonging.berkeley.edu/roots-structural-racism.

Myers, Justin Sean. 2022. *Growing Gardens, Building Power*. New Brunswick, NJ: Rutgers University Press.

Ney, Jeremy. 2022. "Food Deserts and Inequality." Social Policy Data Lab, Malcom Wiener Center for Social Policy. https://www.socialpolicylab.org/post/grow-your-blog-community.

Obama White House. 2010. "Eliminating Food Deserts in America." Promotional Video. 24 February. https://obamaWhitehouse.archives.gov/featured-videos/video/2010/02/24/eliminating-food-deserts-america.

Penniman, Leah. 2018. *Farming While Black: Soul Fire Farm's Practical Guide to Liberation on the Land*. White River Junction, VT: Chelsea Green.

Reese, Ashanté. M. 2018. "'We Will Not Perish; We're Going to Keep Flourishing': Race, Food Access and Geographies of Self-Reliance." *Antipode* 50 (2): 407–424.

Reese, Ashanté. M. 2019. *Black Food Geographies: Race, Self-reliance, and Food Access in Washington, DC*. Chapel Hill: University of North Carolina Press Books.

Robinson, Cedric J. 2000. *Black Marxism: The Making of the Black Radical Tradition*. Chapel Hill: University of North Carolina Press.

Rothstein, Richard. 2017. *The Color of Law: A Forgotten History of How Our Government Segregated America*. New York: Liveright.

Slater, Gene. 2021. *Freedom to Discriminate: How Realtors Conspired to Segregate Housing and Divide America*. Berkeley, CA: Heyday.

Stone, Clarence. 1989. *Regime Politics: Governing Atlanta 1946–1988*. Lawrence: University of Kansas Press.

Tuck, Eve, and K. Wang Yang. 2012. "Decolonization Is Not a Metaphor." *Decolonization Indigeneity, Education and Society* 1 (1): 1–40.

Urban Land Institute. 2018. *St. Paul Minnesota: The Rondo Community Land Bridge*. Washington DC: Urban Land Institute. https://americas.uli.org/rondo-saint-paul-mn-advisory-services-panel/.

Ver Ploeg, Michele, David Nulph, and Ryan Williams. 2011. Mapping Food Deserts in the United States. USDA Economic Research Service. https://www.ers.usda.gov/amber-waves/2011/december/data-feature-mapping-food-deserts-in-the-us/.

Wilkerson, Isabel. 2020. *The Warmth of Other Suns: The Epic Story of America's Great Migration*. London: Penguin Books.

Wolfe, Patrick. 2006. "Settler Colonialism and the Elimination of the Native." *Journal of Genocide Research* 8 (4): 387–409.

Zurawski, Erica. 2023. "Materializing the Metaphor: Theorizing the Food Desert as a Socio-Spatial-Legal Instrument in the Production of Space." *Human Geography* 16 (3): 286–298. https://doi.org/10.1177/19427786231173631.

7 Gentrification, White Privilege, and the Political Shift from Food Deserts to Food Apartheid: Reflections from a Millennial Food Justice Scholar

Justin Sean Myers

It was July of 2006, and my girlfriend, Shay, and I, both white, found ourselves boarding an airplane in Oakland destined for New York City. We were full of excitement, as well as trepidation, about moving to the Big Apple. I had been accepted into the sociology Ph.D. program at the City University of New York, and Shay had gotten a job teaching English at a middle school on the Upper West Side. The prior month, through Craigslist, we had found a two-bedroom, one-bath apartment with a shared backyard in Bushwick, Brooklyn, a working-class Black, Latinx/e, and Caribbean community. The apartment was close to Maria Hernandez Park, not too far from the Central Avenue M subway stop, and a long walk to the Morgan Avenue L subway stop. We thought it would be the perfect place to start our New York adventure.

Upon our arrival we set out to buy some groceries. What we encountered was shocking, to say the least. There were many bodegas full of candy, soda, and snack foods, although a few had delis where you could get a fresh sandwich. Fast food restaurants were also common, from fried chicken to burgers to Chinese. Luckily, we found two grocery stores within a one-mile walk of our apartment too; this was good news to us since we did not own bicycles or cars and were reliant on our feet and public transit to get around. However, they were completely run down with dark lighting and dusty shelves and lacked the produce and food items we were used to. There was no local produce, nut-based milks, fair-trade products, or organic brands. This lack at the two grocery stores was compounded by the absence of farmers markets and consumer-supported agriculture (CSA) programs in the area. Coming from affluent white suburbs in the Bay Area, we were used to being able to shop at numerous upscale grocery stores, not to mention having abundant farmers markets available to us.

The community was indeed full of food, but from our perspective, it was full of all the wrong kinds of food. This realization sent us into a mini-panic as we thought, "What are we going to do for food? We do not want to live off of the food that is close to our apartment." We searched for grocery stores and farmers markets outside the neighborhood and found them—Trader Joe's, Whole Foods, and the Union Square farmers' market. We also began to use the term "food desert" to describe our neighborhood to ourselves, our friends, and our family, for that is what it felt like to us, a place bereft of *good food* options.[1] While shopping outside our neighborhood is how we solved our problem of living in a self-proclaimed food desert, our biggest wish, food-wise, was that a better grocery store would come into our neighborhood, maybe even a farmers market. In short, we wished for "supermarket greenlining" (Anguelovski 2016).

I tell this story of our experience as gentrifiers in Brooklyn because I contend that our collective shock at moving into the food environment of a working-class community of color, the barriers we faced in reproducing our class-based food preferences, and the steps we took to acquire food that was acceptable to us is not an experience unique to us. Rather, it is a shared condition among many first-wave gentrifiers, and unpacking the social processes shaping the first-wave gentrifier provides several insights into the relationship between gentrification, food politics, and the food desert trope (cf. Alkon et al. 2020; Anguelovski 2015, 2016; Hill 2017). Foremost is that how gentrifiers respond to inequitable food relations plays an important role in either reproducing or challenging urbanized racial inequities. And through autoethnographic reflections documenting my own shift from a reactionary supply-side politics to a transformative food justice politics, I hope to underscore how those involved in realizing a more just and sustainable world could use particular concepts, experiences, and frames to push gentrifiers and other privileged people into working with, and not against, movements for racial equity.

The Fears of Millennial Foodie Gentrifiers

While Brooklyn was framed as the "place to be" when we moved to New York, Bushwick was predominantly a working-class community of color; 70.3 percent of residents were Hispanic, 20.6 percent Black, 4.1 percent non-Hispanic white, and 3.2 percent Asian.[2] The vast majority of residents were

Puerto Rican, Dominican, and African American, with growing Ecuadorian and Mexican communities. A total of 32.9 percent of residents lived in poverty, 81.3 percent were renters, over 35 percent were severely rent-burdened (spending over half of their monthly gross income on rent), and the median household income was $33,240 ($51,861 in 2024 dollars).[3] In 2006, gentrification was occurring at the edges of the community along the L subway line, on its northern end as old industrial factory buildings were being turned into lofts and on the southern end along the M, J, Z subway line with the renovation, for gentrifiers, of older apartment buildings and two- and three-family row houses.

We lived outside this initial gentrification zone and were two of the handful of white people on our block. Commercially, Knickerbocker Avenue was the heart of Bushwick and catered to its long-term residents with the Associated and Met grocery stores, bodegas, fast-food restaurants, 99 cent stores, and discount clothing stores. There was one health food store by the L train that catered to the gentrifiers living in the nearby lofts, Brooklyn Natural. The store did sell organic produce, nut-based milks, and organic frozen foods and had a good deli, but it was also fairly small and had higher prices than we were used to.[4] Outside of Brooklyn Natural, the famous foodie scene of Bushwick had yet to establish itself. At this time, it consisted of the farm-to-table restaurant Northeast Kingdom, the cantina at the tortilla factory Tortilleria Mexicana Los Hermanos, the bar The Wreck Room, and the bar and music venue Goodbye Blue Monday. It wasn't until three to five years later that the area had the Bushwick Food Cooperative, the farmers market at Maria Hernandez Park, Roberta's pizzeria, Momo Sushi Shack, Athom Café, and the bars The Narrows, Tandem, and 101 Wilson Avenue.

As previously explained, such a food environment caused stress and anxiety for us, led us to call the community a food desert, and pushed us to shop at Trader Joe's, Whole Foods, and the farmers market near Union Square. In his book *Work, Consumerism and the New Poor* (2005), Zgymunt Bauman claims that "every type of social order produces some visions of the dangers which threaten its identity" (78). These "images of threat" are "projections of a society's own inner ambivalence about its ways and means, about the fashion in which it lives and perpetuates its living" (78). In the case of high-income countries, such as the US, "it is in the image of the poor that we tend to invest our hidden fears and anxieties" (1). These fears and anxieties are not static though; as societies change, so does the image of the poor and the fears and anxieties that are mapped onto them by the middle class and upper class. Bauman states, "It is one thing to be poor in a society of producers and universal employment; it is quite a different thing to be poor in a society of consumers, in which life-projects are built around consumer choice rather than work, professional skills or jobs" (1). For Bauman, we have moved "from a society guided by the work ethic to one ruled by the aesthetic of consumption," and in this new world the haunting specter of the poor that threatens the affluent, and society as a whole, is the status of being a "flawed consumer" rather than "being unemployed" (1–2).

Josee Johnston and Shyon Baumann's book *Foodies: Democracy and Distinction in the Gourmet Foodscape* (2010) builds on Bauman's insights into postindustrial society and its cultural anxieties and status panic by discussing how in an increasingly food-obsessed society the simple "act of choosing food can plunge us into deep anxiety" precisely because food "operate[s] as a source of status and distinction for economic and cultural elites" and is deeply interconnected with "maintaining and reproducing social inequality" (xii, xv). For Johnston and Baumann, the clearest expression of this compulsion is the foodie, a social group whose "interest in food [is] a central part of who they really are . . . [as] status-seeking people" (63, 94), an identity cultivated through food being marked by practices of cultural and economic exclusivity, such as rarity, locality, organicness, hand-craftedness, and authenticness. For the foodie, food takes on more and more importance in one's life alongside the more conventional status objects attached to middle-class and upper-class identity. Consequently, threats to the food-making practices of class identity take on a heightened importance since more energy, time, and money are invested in food as a status-making object.

This exclusivity is rooted not merely in cultural capital and economic capital, a key focus of Johnston and Baumann's work, but geographic access as well. However, high-end grocery stores are generally not part of the first wave of gentrification, which means the gentrifier who is also a foodie (not all are) may encounter the reality that their new neighborhood excludes them from the symbolic and material economy that their racialized class-based identity is dependent on. This creates the problem of how and where the first-wave gentrifier foodie obtains the food that is integral to their race and class privilege? It often means leaving the neighborhood to go grocery shopping elsewhere. I contend that this shared experience among first-wave gentrifiers, the threat of living in a food desert, a threat that is shared with their family, their friends, and the broader white public, is one reason for the frequency with which urban working-class communities of color are labeled as food deserts even when they are not technically so according to the definitions of the USDA or public health agencies—a mislabeling that happens all too frequently (Myers 2023; Taylor and Ard 2015).

When gentrifiers use the language of food desert to refer to the neighborhoods they have moved into, they are not doing so because they are ignorant or unscientific. They are using the term in a manner that makes sense to them, a symbolically and culturally loaded concept that signifies that the area is a space where they are unable to culturally reproduce themselves. The food environment of their new neighborhood is seen as the antithesis of everything they try to craft themselves to be: the food is industrial and mass-produced; it is ultraprocessed and full of chemicals; there is no single-origin sourcing and no unique origin story behind the food to weave a personal connection to the product. Consequently, the food desert, symbolically and materially, manifests itself as the negation of the foodie and compounds anxieties wrapped up with the declining returns to whiteness and a society that is being transformed through escalating rates of inequality, declining rates of social mobility, real-wage stagnation, ballooning amounts of student debt, and unaffordable housing.

Let us step into Shay's and my shoes as first-wave gentrifiers. Our life history up to landing in Bushwick was one where we had generally been able to exercise our freedom and choice in and through the market- and consumption-related practices. We had the economic and geographic access to buy local, organic, and artisanal goods at not just one but several high-end grocery stores. We had access to farmers markets and CSA programs. The racialized capitalist market had worked well for us. We were used to voting with our fork and had not faced the adversity or oppression that requires community organizing, political mobilization, and nonviolent direct action to right inequitable food relations. Additionally, we tended to lack the ability to conceptualize the sociohistorical production of food apartheid in Bushwick by way of segregation, redlining, block busting, planned shrinkage, and the new Jim Crow. Without this quality of mind, we failed to make connections between federally subsidized white flight and supermarket flight and thus the role of the state and capital in producing food apartheid in the community (McClintock 2011; Myers 2023; Valle and Guthey 2023; for more on food apartheid see chapter 6). Moreover, we had been raised in a society where food activism had been driven by and shaped by a narrative where (individualized) consumerism was seen as the only effective form of political action (Szasz 2007).

So when first-wave gentrifiers, like Shay and me, encounter a neighborhood with limited food access, they by and large, do not fight for demand-side policies and programs to raise the incomes of long-term residents and do not work in solidarity with existing community-based efforts rooted in self-determination instead, they either leave the community to procure good food for themselves, demand that better stores locate in the community, or team up with others like themselves to create alternative consumer options to the conventional choice (e.g., a buying club or food cooperative that is viewed by long-term residents as a "white space" and thus not frequented by them) (cf. Anguelovski 2015, 2016; Zitcer 2015).[5] Given the weak ties between gentrifiers and long-term residents and the ways of thought and the politics that gentrifiers bring with them, another outcome is that gentrifiers often do not concern themselves with whether long-term residents want high-end stores or how the in-movement of such stores would impact long-term residents—whether they could afford the food at such stores and how such stores might drive up residential and commercial rents and thereby displace long-term residents and the existing food establishments catering to them (Anguelovski 2015, 2016).

Such actions, while harmful to long-term residents, are also reflective of what Johnston and Baumann call the "consumer-ethics frame," which is the guiding logic of foodies and the food movement in general, one where "great weight [is] given to market retailers to enable and influence these [good food] decisions" (2010, 165). Given the centrality of political consumerism and good food retailers to foodies and the food movement, it is not too far a leap to grasp how public-private initiatives emerge to incentivize grocery stores to come back to urban working-class communities of color and why these initiatives become the key policy program for addressing grocery store flight (Chrisinger 2016; Doshna 2015; Karpyn et al. 2010; Wolf-Powers 2017; Zurawski 2023). Such programs are techno-fixes to a social problem that emerges from the cultural class interests of middle-class and upper-class white folks, one that easily maps onto the hegemony of neoliberal supply-side politics—presuming that consumers have income to spend but not stores to spend it at. The popularity of the food desert trope is therefore due in part to its ability to assist an increasingly urbanizing white America in understanding the poverty and inequality that now, due to gentrification, seemingly lives all around them (rather than such poverty and inequality being spatially segregated from suburbia and therefore out of view and out of mind except as mediated though the mass media).

In avoiding the racialized roots of inequitable food relations, the food desert trope focuses on inequities in consumption, not production and distribution or the relationship between the three. The food desert trope also ignores that bringing a grocery store, particularly a high-end store, to a gentrifying community without addressing a society structured around low-road jobs, the evisceration of the social safety net, mass incarceration, fiscal austerity, and trickle-down economics will primarily benefit white gentrifiers not long-term working-class residents of color (Myers and Caruso 2016). At the same time, the food desert trope is widely popular precisely because its ahistoricity and geographic-centricity do not challenge white people's racialized class interests or the broader structures of white supremacy and therefore do not require the first-wave gentrifier to question their position and role in sociohistorical processes and structures producing rising inequality, declining social mobility, and the consolidation of white privilege. Consequently, the ascendancy of the food desert trope is directly connected to its ability to protect and shield the racialized class interests and positionalities of the gentrifier and their desire for supermarket greenlining.

In this respect, echoing Hill (2017), I contend that the food desert trope operates as a "controlling image" because it is produced and circulated by and for white America; it expresses the anxieties and fears that white America has about itself at a time when it is displacing residents of color from their longtime communities; and it performs the ideological work that maintains existing racial and class inequities while masking the political, economic, and social structures that produce inequitable food relations (cf. Collins 2000). The food desert trope subsequently underscores how the oppression of communities of color is reproduced through ideological processes that legitimate material domination, for the symbolism and framing of the food desert trope shapes how white America sees the problem *of* and the solution *to* food deserts, including whether they can be addressed, to what extent they can be addressed, and how they might be addressed, all in ways that benefit white America and racial capitalism.

Transforming the Gentrifier: From Reactionary Politics to Solidarity Politics

While our initial reactions to the food environment in Bushwick maintained and amplified class and race inequities, I eventually rejected the food desert trope, political consumerism, and supply-side politics in favor of a food justice politics rooted in procedural justice, one that challenges white supremacy, pushes white people to interrogate the social roots of inequitable food relations, and prioritizes working in solidarity with the existing efforts of residents of color.

Three experiences were integral to this transformation. The first was an environmental justice course I took as a graduate student. Reading books on the environmental justice and food justice movements, and learning about their political strategies and ideological frames, pushed me to see how distributive inequities were primarily due to political and economic marginalization and that solutions cannot be technocratic, top-down, and externally led, for such solutions only reinforce the conditions generating the inequities in the first place. Based on these insights, I realized food inequities would not be ameliorated, let alone eliminated, without a change in political power between ethnoracial and class groupings, a significant transformation in the decision-making structures of governance, and an asset-based and self-determination-based development model. Central to this change in thinking was the concept of *just sustainabilities*, which place justice, equity, and redistributive politics at the core of environmental, food, and sustainability politics (Agyeman 2005). This framework, and its privileging of procedural equity as the pathway toward substantive and distributive equity, began to guide my thinking about how to create a more just and sustainable society.

Such a framework pushed me to work in solidarity with the efforts of long-term residents to address inequitable food relations rather than put forth gentrifier-led ideas. This guided me to my second experience, my dissertation project with East New York Farms! (ENYF!), a food justice organization in the working-class Black, Caribbean, and Latinx/e community of East New York, Brooklyn (Myers 2023). ENYF! emerged in 1998 out of an asset-based participatory planning project that sought to address numerous social, economic, and ecological inequities facing residents: food insecurity, youth unemployment, and land tenure. Today, the organization operates two urban farms, two community gardens, two farmers markets, and a thirty-member youth program, and partners with over forty resident-run food-producing community gardens.

In my time working with residents and ENYF! staff, I learned that the food inequities the community faced—being underserved by grocery stores, having low-quality grocery stores, having way more bodegas and fast-food restaurants than grocery stores, and experiencing high rates of food insecurity, diabetes, and hypertension—were not the result of market failures or the personal choices of residents; they were due to a history of political and economic marginalization and authoritarian urban planning rooted in anti-Blackness, anti-Latinxness, and systemic racism. I also became aware of the various strategies of resistance that residents utilized to address the food inequities in East New York, including turning vacant lots into community gardens, protesting the poor conditions of local grocery stores, protesting against Walmart as a solution to the community's food inequities, shopping outside the community, trying to get bodegas to sell produce, starting CSAs, and advocating for the creation of high-road jobs and affordable housing in the community. These encounters with grassroots organizations and bottom-up planning models pushed me to think critically about how solutions cannot be limited to food and the food system but must address the broader social system generating food inequities. This meant that I needed to actively support the calls of local organizations for the democratization and decentralization of political power to residents through community-based planning processes, participatory budgeting processes, community land trusts, and community-owned housing cooperatives in order to create the conditions for equity and self-determination for communities of color (for more on such efforts see chapters 8, 13, and 14).

Even with these experiences, I was lacking the language and framing that succinctly challenged the discourse of food deserts and its ahistorical, geographic-centric, and neoliberal supply-side politics, especially when it came to talking to other privileged white people about what is producing inequitable food relations. The inability to name the problem remained with me until I heard two Black women in the New York food justice movement, Leah Penniman and Karen Washington, use the term *food apartheid* at local food conferences to describe the social, economic, political, and racial roots of inequitable food relations.[6] In articulating the language of food apartheid rather than food desert, activists such as Penniman and Washington push white people like myself to move away from the naturalism, environmental determinism, and ahistoricism of the food desert trope and toward wrestling with the deep sociohistorical processes that have denied communities of color the income, wealth, land, education, and decision-making power necessary to live full lives.

Food apartheid is not only a more accurate way of framing inequitable food relations, but it also employs a language that directly connects food inequities to systemic racism, anti-Blackness, and anti-Latinxness in a manner that the food desert trope is incapable of. In naming the problem as one of systemic racism and white supremacy, food apartheid points to very different solutions than those attached to the food desert trope and its politics of subsidizing the costs of for-profit grocery stores. By invoking segregation and systemic oppression, food apartheid emphasizes that solutions need to be about more than food; solutions need to entail the democratization of power and the creation of policies and programs that address the marginalization of and lack of self-determination for communities of color.

Transforming the Gentrifier

In 2006 Shay and I became first-wave gentrifiers; while we initially maintained our individualistic voting with your fork politics as the solution to the geographic access barriers we encountered, my subsequent experiences reading environmental justice scholarship, working with the food justice organization ENYF!, and being introduced to the language of food apartheid were all central to moving beyond our demand for supermarket greenlining as sufficient to addressing food inequities. Yet such experiences also underscore that this transformation required that I challenge my own privileged positionality with respect to race and class, interrogate the social roots of inequitable food relations, attempt to work in solidarity with the existing efforts of residents of color, and adopt terminology that would challenge how I and other white people conceptualize the roots of and solutions to inequitable food relations. My hope is that in sharing this story, you are able to glean some ideas about how to transform the politics of the gentrifier so that they don't double-down on supermarket greenlining but, rather, challenge food apartheid and, in so doing, work in solidarity with the efforts of communities of color for self-determination.

Notes

1. I use the term "good food" based on how the middle-class and upper-class white gentrifier socially constructs its meaning through their racialized class position. In doing so I underscore that this term is not connected in any way to notions of physiological health or an "objective" scientific definition of healthy food.

2. Information based on author's calculations, American Community Survey (ACS) 2005–2007. Retrieved October 22, 2018, https://data.census.gov/.

3. Information based on the author's calculations. American Community Survey (ACS) 2005–2007, retrieved October 22, 2018, https://data.census.gov/.

4. Complaints of high pricing around this time are common on Yelp reviews for Brooklyn Natural; see https://www.yelp.com/biz/brooklyn-natural-brooklyn.

5. This summation is also informed by my informal conversations with first-wave gentrifiers in Brooklyn, including those who were members of the Bushwick Food Cooperative, as well as those involved in the food movement in Pittsburgh.

6. Karen Washington is credited with coining the term "food apartheid" and is Leah Penniman's mentor. See "How to Grow Change Through Black-Led Agriculture, According to Leah Penniman," *Food & Wine*, June 14, 2020.

References

Agyeman, Julian. *Sustainable Communities and the Challenge of Environmental Justice*. New York: NYU Press, 2005.

Alkon, Alison Hope, Yuki Kato, and Joshua Sbicca, eds. *A Recipe for Gentrification: Food, Power, and Resistance in the City*. New York: NYU Press, 2020.

Anguelovski, Isabelle. "Alternative Food Provision Conflicts in Cities: Contesting Food Privilege, Injustice, and Whiteness in Jamaica Plain, Boston." *Geoforum* 58 (2015): 184–194, https://doi.org/10.1016/j.geoforum.2014.10.014.

Anguelovski, Isabelle. "Healthy Food Stores, Greenlining and Food Gentrification: Contesting New Forms of Privilege, Displacement and Locally Unwanted Land Uses in Racially Mixed Neighborhoods." *International Journal of Urban and Regional Research* 39, no. 6 (2016): 1209–1230.

Bauman, Zgymunt. *Work, Consumerism and the New Poor*. New York: Open University Press, 2005.

Chrisinger, Benjamin W. "Taking Stock of New Supermarkets in Food Deserts: Patterns in Development, Financing, and Health Promotion." *Center for Community Development Investments*, 2016. https://www.ncbi.nlm.nih.gov/pmc/articles/PMC6296768/.

Collins, Patricia Hill. *Black Feminist Thought: Knowledge, Consciousness, and the Politics of Empowerment*. New York: Routledge, 2000.

Doshna, Jeffrey Peter. "Community Development in the Age of Neoliberalism: The case of the Pennsylvania Fresh Food Financing Initiative." PhD dissertation, Rutgers University, 2015.

Hill, Alex B. "Critical Inquiry into Detroit's 'Food Desert' Metaphor." *Food and Foodways* 25, no. 3 (2017): 228–246, https://doi.org/10.1080/07409710.2017.1348112.

Johnston, Josee, and Shyon Baumann. *Foodies: Democracy and Distinction in the Gourmet Foodscape*. New York: Routledge, 2010.

Karpyn, Allison, Miriam Manon, Sarah Treuhaft, Tracy Giang, Caroline Harries, and Kate McCoubrey. "Policy Solutions to the 'Grocery Gap.'" *Health Affairs* 29, no. 3 (2010): 473–480.

McClintock, Nathan. "From Industrial Garden to Food Desert: Demarcated Devaluation in the Flatlands of Oakland, California." In *Cultivating Food Justice*, edited by Alison Hope Alkon and Julian Agyeman, 89–120. Cambridge, MA: MIT Press, 2011.

Myers, Justin. *Growing Gardens, Building Power: Food Justice and Urban Agriculture in Brooklyn*. New Brunswick, NJ: Rutgers University Press, 2023.

Myers, Justin Sean, and Christine C. Caruso. "Towards a Public Food Infrastructure: Closing the Food Gap Through State-Run Grocery Stores." *Geoforum* 72 (2016): 30–33.

Szasz, Andrew. *Shopping Our Way to Safety: How We Changed from Protecting the Environment to Protecting Ourselves*. Minneapolis: University of Minnesota Press, 2007.

Taylor, Dorceta E., and Kerry J. Ard. "Food Availability and the Food Desert Frame in Detroit: An Overview of the City's Food System." *Environmental Practice* 17, no. 2 (2015): 102–133.

Valle, Gabriel, and Greig Tor Guthey. *Food Apartheid in California and Beyond*. Report for the Center for California Studies, CSU Sacramento, 2023.

Wolf-Powers, Laura. "Food Deserts and Real-Estate-Led Social Policy." *International Journal of Urban and Regional Research* 41, no. 3 (2017): 414–425.

Zitcer, Andrew. "Food Co-ops and the Paradox of Exclusivity." *Antipode* 47, no. 3 (2015): 812–828.

Zurawski, Erica. 2023. "Materializing the Metaphor: Theorizing the Food Desert as a Socio-Spatial-Legal Instrument in the Production of Space." *Human Geography* 16, no. 3 (2023): 286–298.

8 No Justice Without Land: The Struggle for Autonomy in the San Diego Urban Agriculture Movement

Belinda C. Ramírez

> We should not only be raising our gardens, we should be raising a ruckus!
> —Robin Wall Kimmerer,[1] Potawatomi botanist and author

Introduction

In popular media, urban agriculture is often characterized as a leisurely activity for those with excess time and money. However, in many neighborhoods within and beyond the US, it is informed by visions of autonomy and self-dependence that can serve more politically radical and racially resilient endeavors. In these cases, community practices of urban agriculture operate as political pushback to racial and economic inequalities, and these resistance politics affect the social and collective values of food producers in urban neighborhoods (Cockrall-King 2012; Dawson and Morales 2016; Myers 2023; Premat 2012; White 2011a, 2011b). Urban agriculture serves as a site for the act of place-making, for community-making, and for urban food growers to negotiate personal and collective value and meaning (Hammelman 2022; Raja et al. 2024). Value, community, and place—and the relationship between these concepts—therefore deeply affect efforts at self-determination for people of color within the urban agriculture movement.

However, one of the most difficult challenges faced by urban agriculture practitioners—especially practitioners of color and those with few financial resources—is access to land. "Sovereignty implies 'space,'" as French philosopher Henri Lefebvre states, and as such, the struggle for autonomy entails a deep engagement with the spatiality and territoriality of that struggle (Lefebvre 1991, 280; Gilmore 2017). I therefore argue that urban agriculture is an important form of place-making that has radical political potential, especially in communities of color. However, I will also demonstrate that this potentiality is constrained by opposing forms of land valuation that occur under neoliberal capitalist logics, which threaten food-growing practitioners' rights to space and the city.

In this chapter, I draw from thirty-six months of deeply participatory ethnographic fieldwork—as a queer, Latine urban farmer myself—as well as interviews and archival work, to provide examples of how tensions and confrontations between place-making and capitalist development play out in the lives of BIPOC community gardeners in the San Diego–Tijuana area. This work was greatly influenced by grounded theory, allowing what was happening on the ground to shape my research questions. The first case study comes from a Latine-run community garden in National City, and the second comes from a Black-led community garden in southeastern San Diego. In each of these, La Vía Campesina's statement of "queremos una tierra para vivir" ("we want land to live") comes to life, but the cases also demonstrate the difficulty in achieving autonomy and sovereignty within communities' lived realities. Those involved in the urban agriculture movement often engage with the question of how to create something that does not already exist within the confines of what *does* already exist. So, the challenge presented to these sustainability-minded dreamers is this: What practical, tangible steps could be taken to move their ideas from thoughts and aspirations to a new reality, a new way of being, even a new consciousness and moral code to confront neoliberal logics? Throughout my fieldwork, I observed the conflicts and tensions around this question that were deeply entangled with racism, classism, sexism, and other power differentials.

The Role of Land and Place-Making in Food Justice

Within cityscapes, land, space, and property are highly sought after and charged with meaning in a multiplicity of ways: space and land as capital, as private property, as solace and home, as community and identity (Fainstein 2001; Hackworth 2007; LaDuke 1999; Shiva 2014). However, not all of these meanings hold equal weight in modern, Western society. As Holt-Giménez and Williams (2017) write, "Our skewed systems of land tenure reflects [*sic*] a regressive political-economic system, itself embedded in a continuing legacy of dispossession, concentration and exploitation," and virtually no space is left untouched by such forces (259). This has only increased since the time of their writing, as land ownership is even more concentrated and financialized today, becoming "one of the major forms in which wealth is stored" (Fairbairn 2021, 12; see also Clapp and Isakson 2018; Ross 2014; and Rosset et al. 2006).

But food-growing practitioners in urban settings are trying to tip the scale in a different direction, because more than a wealth-generating mechanism, "land is a crucially important source of food and livelihoods" (Fairbairn 2021, 12). As any urban farmer or gardener would tell you, land is a central component in urban agriculture and yet also one of its biggest challenges. And land is not central merely because you need land to grow food. Because the formation of community around urban agriculture takes place through events and stories, conflict and recovery, connection and hope, land becomes imbued with meanings, which is both a social and a physical process (see, for instance, Birkhold et al. 2014; McClintock 2018, 2021; Premat 2012; White 2011a, 2011b; and see chapters 1, 4, and 15 in this volume, which all discuss various racial, ethnic, and political influences within food justice activism and how those influences shape the ways urban agriculture and land are interpreted and prioritized, or not).

According to Marxist analyses, the city is a privileged space for the consumption of commodities, and the city itself is consumed as if it were one large commodity. The city was borne of industrialization, which produces urbanization. In fact, according to Lefebvre, the epitome of capitalist production is lived in the lives of "the urban"—or those that dwell in urban spaces (Lefebvre 1996). Many have even argued that urban environments are the primary tool of expansion for the neoliberal political project, since they provide the capital, consumption, and people necessary to prop up such a system (Eizenberg 2012; Harvey 1985; Hackworth 2007; Pinson and Journal 2016; Purcell and Tyman 2019). Because of this, David Harvey and Henri Lefebvre talk about the "right to the city," which is a collective rather than an individual right, a kind of proletariat reclaiming of spatiality and power. It is collective because of the shared power needed to transform and reshape the processes of urbanization (Harvey 1996, 2008; Lefebvre 2003).

In this vein, the urban agriculture movement attempts to reframe the story of capital and hegemony in order to create new knowledges. In an effort to compete against large social, political, and economic regimes, urban agriculture endeavors to sustain an alternative model of being against the prevailing ones, thereby shaping and cultivating a new world outside the realm of capitalism and development. Holt-Giménez and Williams say it succinctly: "Just as we can't isolate food justice from land justice, we can't isolate land justice from the need for deep political and economic transformation" (2017, 259). Urban agriculturalists are attempting to do just that.

In a neoliberal context, the physical, spatial sites of urban agriculture are also explicit efforts at reclaiming the commons, which demonstrates the constant negotiation of the production of urban space between several actors—the public, civil society and state and municipal governments (Eizenberg 2012; Harvey 1996; Linebaugh 2014; McClintock 2021; Shiva 2020; White 2019). And within this environment there is, of course, conflict and contestation. Despite the dialectic nature of urban agriculture sites—that is, being simultaneously neoliberal and radical—they *are* part of a wider phenomenon of urban contestation by which space is utilized to voice and fight for alternative sociopolitical arrangements (Cutts et al. 2017; McClintock 2014). These places and the land they inhabit are suffused with experiential qualities, which provides a grounding for sociopolitical projects (Casey 1998). Place is a constitutive feature of human habitation, and defining the contours of the production of space allows us to specify how a politics of food is spatialized and how relations, practices, and value or meaning are produced, evaluated, and contested via place (Premat 2012). The case studies that follow will demonstrate what happens when these radical place-making practices encounter the concentrated and racialized politics of land.

Mundo Gardens and Joe's Pocket Farm

Mundo Gardens is a nonprofit based in National City and led largely by the Reynosos, a Latine family. National City has a population of around 63,000 with 63 percent of that population identifying as Hispanic and 20 percent as Asian and Pacific Islander, mostly Filipino. Due to long histories of racial segregation through practices such as redlining and the void of social services (see chapter 6 of this volume, which discusses these kinds of exclusionary practices in depth), National City has been challenged by the prevalence of chronic disease among its residents. In fact, the rates of diabetes hospitalization and mortality are 2.3 times higher in National City than other areas of San Diego (CHIP 2018). The city has acknowledged these issues and their connection to unhealthy land use practices, leading them to adopt a health and environmental justice element in their general plan in 2011. They were the first city in California to do this, though polluting industries have not been removed from the area (CEJA 2021).[2]

Mundo Gardens had been informally managing an urban agriculture site called Joe's Pocket Farm for over ten years. The site was named after José Núñez, who grew crops there nearly thirty years ago (Hernández 2018a, 2018b). After he became ill and moved, the site became derelict until the Reynoso family and others decided to revitalize it and name it in honor of its original founder. Over the years, the farm became a community hub, with Mundo Gardens holding regular events, volunteer work days, and agricultural education opportunities on the site. In 2017, the Reynosos applied for a temporary event permit on the site, which this time prompted city officials to investigate whether the land was city property. Finding that it was, the city denied the permit until management of the farm could be established. The Reynosos took these efforts to City Hall, gaining support for the project from city officials.

The subject of Mundo Garden's official management of Joe's Pocket Farm was finally determined at the July 2018 city council meeting, which I attended. The chambers held around thirty supporters—nearly all Latine—holding small signs showing their support for Mundo Gardens. Nearly all of these supporters also gave public comments, sharing their belief with council members and others in attendance that Mundo Gardens had helped provide fresh and healthy food to their children and had bolstered and created community. A white woman I recognized as Stepheni Norton, the owner of Dickinson Farm—a for-profit farm also located in National City—and her husband Mike, also white, were in the crowd, too. Norton went up to the podium to share her thoughts. After an introduction that stated she was in support of Mundo Gardens, as well as Olivewood Gardens and Brightside (another urban garden and a fresh food distribution effort in National City, respectively), Norton's three-minute comments took a turn I did not expect: she declared her opposition to the recommendations made by the city staff, urging Mundo Gardens, OliveWood, and Brightside to instead "get the appropriate licenses." Her remarks shifted toward emphasizing "rules," with her saying that word repeatedly throughout the talk. She mentioned that "some follow the rules and some talk to staff and get what they want," obviously alluding to herself, and her business, Dickinson Farm, as a rule-follower and the Reynoso family, with Mundo Gardens, as those who circumvent rules and do as they please.

In a later NBC 7 news segment that covered this controversy, Norton was highlighted as saying, "They are the rules and all of us have to follow them. And when people aren't following them, it's not only discriminatory by the agency that's allowing it, it's also potentially not safe" (Little 2018). Norton's invocation of rules is racialized, given that communities of color are overpoliced and often perceived as lawbreakers. Lawfulness becomes an exclusionary tool and a weapon in the fight for the white status quo, especially given the contentious politics surrounding immigration in this borderland region (Brettell and Nibbs 2010; see also Chavez 2013).

Norton's comments seemed especially odd and contradictory given that she had made supportive comments on a 2018 article in the *San Diego Reader* that praised Mundo Gardens for taking the lead on the Paradise Creek Park Community Garden. Norton wrote, "We are so excited about the new community garden! With Joe's pocket garden, Olivewood Gardens donation-based farm stand and our CSA and weekly farm stand, the fresh and healthy National City grown options are available for community members regardless of what part of National City they live in. #bethechange" (Gavidor 2018). What had shifted Norton's opinion in the intervening six months is uncertain.

Over the year that I had been acquainted with Norton, she consistently emphasized how it took her eighteen months to get legal permission to operate her farm. It felt odd for her to keep bringing it up and stressing the time it took to achieve these permissions. But this context clarified her comments at the City Council meeting and the underlying values, unspoken and unacknowledged privilege, and obscured race- and class-based critiques behind them. Her husband, Mike, was the last to speak during public comments, reiterating Norton's points and introducing his own brand of concerns and mysteries, including having the soil at Joe's Pocket Farm tested in case it contained toxic elements. Mike and Stepheni's comments seemed to make a big impact on three of the five council members, with subsequent deliberation moving into issues of liability, possible soil contamination, rules and regulations, and allusions to a lack of competency on the part of Mundo Gardens.

The once humble and sensitive crowd that had come to support Mundo Gardens was now up in arms, shouting antagonisms at the council members. Ultimately, the staff recommendations were split; Mundo Gardens was granted manager status of Joe's Pocket Farm, but they were not given the same status for another site they were running, El Toyon Farm, or the new and planned Paradise Creek Park Community Garden. Instead, a purportedly more democratic process of a request for proposals and request for quotations would be issued so that other organizations could apply for management of these urban agriculture sites, although according to the Mundo Gardens supporters, such a process had already occurred with no other takers. With the agenda moving forward after the votes had been cast at about half an hour past midnight, the Mundo Garden allies left the Chambers, shouting humor-ridden jabs at the mayor and mumbling begrudgingly. Stepheni and Mike remained in their seats.

Seven months after that night at the National City city hall, I drove by Joe's Pocket Farm and found the space in disarray. The soil tests the city had demanded be administered came back indicating high levels of chlordane and dieldrin, both of which are human-made insecticides found commonly in residential and agricultural spaces due to their prolific use as pesticides between the 1940s and 1970s. In visiting the small farm, I saw that in reaction to this news, the city had restricted use of the garden by erecting a tall chain-link fence around the space and posting signs indicating that the soil was contaminated and food from the space should not be consumed. Newspaper articles quoted Norton as saying she would never eat anything that came from that garden, and the mayor at the time even referred to the garden as an "outlaw-type situation," once again very clearly linking this people of color–led initiative with lawlessness and rule-breaking. Joe's Pocket Farm was effectively dead.

Essentially, the community Joe's Pocket Farm was serving—largely working-class Latine folks in National City—was being punished for living in urban spaces with high levels of contaminants due to their proximity to freeways and decades-old practices of using carcinogenic substances as insecticides. And the punishment? No fresh food. No ability to grow fresh food in community with other like-minded people.

But the Reynosos and other Mundo Gardens supporters are determined. Between 2018 and 2019, they held fundraisers to help cover increased insurance costs, water, and the installation of city-mandated raised beds at Joe's. In 2019, after working with a new Latine mayor (previously one of two councilmembers in support of Mundo Gardens in the 2018 city hall event described above) as well as creating partnerships with well-respected urban agriculture organizations in National City to fulfill the request for proposals and request for quotations requests, Mundo Gardens was approved as the sole operator of the nearby Paradise Creek Community Garden. However, in the summer of 2020 the City of National City made it known that they wanted to sell the land Joe's Pocket Farm is located on without having consulted Mundo Gardens. Protests for the farm emerged simultaneously with the wave of Black Lives Matter activism that followed the murder of George Floyd, gathering dozens of people from all across San Diego to demonstrate that Joe's Pocket Farm was worth saving and worth restoring to community use. As of the time of this writing, no further action has been taken by the city to sell the land, but in February 2025 the city unexpectedly cut down the fruit orchard and trees in the space, including a beloved and historic mulberry tree. The site remains contested and closed to the public.

Project New Village and Mt. Hope Community Garden

Project New Village is a Black-led nonprofit situated in southeastern San Diego that also experienced a recent threat to their land tenure. However, they were able to mobilize resources and respond very differently than Mundo Gardens. When the City of San Diego moved to sell the parcel on which their community garden had been grown, they worked with local philanthropic groups, businesses, food justice advocates, and banks to buy the parcel. And yet despite this seemingly more positive outcome, the capitalist valuation of land also threatens their work in myriad ways.

Southeastern San Diego is known locally as one of the most socially and economically disadvantaged areas in the county, as well as one of the most violent. The neighborhood surrounding Mt. Hope has a population of 60,000 people and is a largely Latine (70 percent), African, and African American community. The median income in the area is just under $40,000, which is just above half the county's median income. Almost a third of the population lives below the federal poverty line, and 21 percent of households in the area receive SNAP (food stamp) assistance. This region was declared a food desert in 2013. And with these statistics come high rates of chronic disease and childhood obesity, just like in National City.

Project New Village (PNV) wants to change the narrative in this area from one of lack and hardship to one of resilience and self-determination. The organization was started in 1994 by N. Diane Moss as a hub for collaborative efforts to increase social and community wellness. In 2008, the organization started its journey to address inequities in the local food system. After Diane

Figure 8.1
A work party at Mt. Hope Community Garden, run by Project New Village. Photo by Belinda C. Ramírez.

and others succeeded in changing local legislation that determined where community gardens could be located, PNV established the Mt. Hope Community Garden in 2011 (see figure 8.1). Many other community gardens popped up in the area around 2010 due to this legislation as well.

PNV also recently expanded its geographical and visionary purview to include the surrounding areas of Lemon Grove, Barrio Logan, and National City. They informally dubbed this the "Corridor of Color," since it boasts one of the largest concentrations of people of color in San Diego County. According to the PNV website, they want southeastern San Diego to be "a community of active neighborhoods supporting and contributing to the health, wealth and well-being of community members."

When Mt. Hope Community Garden began, PNV had an agreement with the City of San Diego to use city-owned land. But early in 2018 a sign appeared at the garden indicating that the land was up for sale. Diane describes what happened next:

> Well, I call this a "minor miracle," that I'll share with you. It's a minor miracle! So you know, this property was city property. One day we came into the garden, there was a big sale sign up there on this property. So we are a grassroots organization, we don't have deep pockets. We saw a sale sign, and we started thinking, "Well, where else can we go to do this [have a community garden]? Because somebody's going to buy it." But we have a cross-sector group of partners, we meet every month. And we're creating something called the "Good Food District." So one of the partners was a developer and a planner, and he said "Let's go for it. Let's bid." We've been working together for ten years, so I trusted him, and this was his area of expertise. So we did what he said. We put in a bid, we outbid three different regular bidders, housing developers, and we won.

The sale of the plot by the city set Diane and her team on the road toward developing a proposal to acquire the land. After an arduous bidding process, they eventually won the bid at the city council and established escrow thanks to various funders who decided to support the project. After the $20,000 down payment to the city, PNV needed to fundraise and acquire financing for a bit more than $600,000 to acquire the land. Diane continues:

> And then another minor miracle happened—so it's a series of miracles here. And this is developing the idea of the Good Food District and this Hub, or this center place for food distribution and food activity. But the second thing is that a national nonprofit, the Conservation Fund, had started coming to our collaborative meetings. They could see what we were aiming to do, that we were sincere, that we had a plan, and they made an offer to us. They said, "If you can come up with $100,000 in pledges and cash, we will loan you the money you need to make the purchase of this land." And we had 60 days to do it! [And] We did it! So now we're in debt, but we own the land under the garden. And we got big, big plans for that.

The Conservation Fund, a white-led environmental nonprofit, helped support what this community-based group was already doing and gave Project New Village an incentivized loan on the condition that they could raise $100,000 in cash and pledges. Through dedicated organizing, drawing on relationships with community partners, and appealing to local philanthropic entities, the small nonprofit achieved this goal and was able to cover the larger sum for land acquisition through the Conservation Fund loan (Brandeis

2019). This was all made possible because of the support network of important individuals (mostly making up the PNV advisory board) that Diane had cultivated over the years and the belief they had in PNV's work in southeastern San Diego. This group was made up of both leaders of other organizations in the area as well as representatives from nearby universities, city departments, and nonprofits. Through her years of organizing, Diane knew how to connect to those who could shape and move her vision forward, while also attaching important institutional power to PNV's work. Now, having celebrated being the owners of the land (with the Conservation Fund as lien holders, essentially) since early 2020, they have turned their sights toward other options to fund the very expensive development of the site.

This development, dubbed the "The Village" (previously referred to as "The Good Food District Hub"), is planned to feature 20,000 square feet of commercial space including a local health food market and prepared local food area, as well as office space for holistic healing, a community kitchen to help cultivate local food entrepreneurs, and 15,000 square feet for low-income senior housing. Elements of urban agriculture will also remain, although the existing garden will need to be relocated.

As the PNV board sees it, The Village addresses the structural, systemic underpinnings of racial inequities. As long-term neighborhood resident and late board member Robert Tambuzi put it, "We need to call racism for what it is. We need to get to a point where people talk about it. The good food system is a point of entry for discussions about institutional racism." The Village is planned to be a space to encourage these kinds of conversations.

This case shows what is possible when a nonprofit rooted in a systematically marginalized community works with wealthier allies to navigate capitalist land markets. But there is worry that this perceived success will compromise their ability to make place while also maintaining their values. For example, while I served on the advisory board, the members engaged in some discussions about how to avoid inviting gentrification to the area through both their urban agricultural initiatives and The Village. Conversations about the local healthy food market, in particular, raised concerns about who the market will cater to and whether those living in the area will be interested in shopping there or even be able to afford it. "How can this work? Is this possible?" advisory members have intoned. But I found that these conversations were never satisfactorily addressed and were often avoided once brought up. During a funders meeting in August 2018, for instance, an academic advisory member argued that since this is a community project, planners should focus on the market: "outsiders will come if neighborhood folks come." But rather than an admonishment against having outsiders to the neighborhood use the market, this was said in a way that was encouraging—the market would need outsiders to shop there in order for it to survive financially. As long as there were neighborhood shoppers as well, though, anxieties around the potentially gentrifying elements of the project could be soothed. Similarly, a couple of months later during a planning meeting, one PhD student advisory board member warned about the perils of gentrification. She said that the Hub "should be for the people in the area," but conceded that "if it brings in some outsiders, too, that's okay." Again, these topics seemed to be actively avoided due to their complicated nature. The financial viability and sustainability of the development hinges on the economic realities of making enough money through sales at the market and prepped food areas and through rent. By necessity, this means that customers need to come from both within the community and outside it in order to generate enough income. These discussions have not led to easy answers or conclusions, but the tension between "staying true" to their values and moving forward with an inherently capitalist, potentially gentrifying project, is always present.

So, despite their initial radical orientations, Project New Village is faced with the economic realities of living and working within neoliberal capitalist structures with respect to organizational needs in governance and fundraising, and ongoing city and business efforts to redevelop and gentrify these majority–Black and Latine areas. Project New Village opted to resolve ongoing threats of eviction by buying the land. In doing so, some radical values were compromised as a sacrifice for the longevity of the organization, and landownership and fundraising became more clearly central to their work, signaling a particular stance on capital, land, and relationships to land that emphasizes colonial and extractive logics.

Conclusion

Lefebvre writes that urbanization is central to the survival of capitalism and is therefore bound to be a crucial focus of political and class struggle. This struggle, he said, would be for the right to the city, or in other words, for the right to command the whole urban process (2003). The work presented in this chapter intends to expand this argument, showing that place- and values-based community practices of urban agriculture therefore operate as political pushback against neoliberal inequality. As has been discussed in other work describing urban agricultural projects around the world, here at the San Diego–Tijuana border urban agriculture attempts to radically reenvision what it means to be urban by injecting place-based values into the pursuit of capitalist development. Urban agriculture, as one leg of the food movement, is an attempt to break down neoliberal urbanism and enact hope and the desire for an alternative way of being in the world, outside neoliberal capitalism and infused with collective, sustainable values. Joe's Pocket Farm and Mt. Hope Community Garden are excellent examples of this attempt, but they also demonstrate how the power differential is very much tilted in the other direction. It is a sign of incredible resilience for organizations like Mundo Gardens and Project New Village and the people who comprise them to continue to push back against hegemonic forces despite the odds being against them and to work within these structures when need be. Such pushback is necessary because these hegemonic power structures are not meant to include poor folks and people of color but instead rely on the continual exploitation of minoritized communities to continue to thrive. If those participating in urban agricultural projects and spaces do not contest those power structures, they are likely to never be heard from and continue to live with the knowledge of their own exploitation. Such a way of living is not justice-based and is not tenable. So back to the soil they go, working the land, composting food scraps, planting seeds, and educating others on what it means to live with respect, sovereignty, autonomy, and love for the land.

Notes

1. Cited in Jewell 2019.

2. One example of such polluting industries comes from an experience during my dissertation fieldwork. On July 12, 2020, I was unable to go to an urban farm where I was working as the culinary gardener because the *USS Bonhomme Richard*, a naval ship stationed in National City's port, caught fire. A tremendous amount of toxic smoke darkened the sky in National City and beyond for several days, making it unsafe for essential workers like farmers to work outside. It took until July 16 for the fire to be fully extinguished.

References

Birkhold, Matthew, Grace Lee Boggs, Rick Feldman, and Shea Howell. 2014. "A Detroit Story: Ideas Whose Time Has Come." In *Grabbing Back: Essays Against the Global Land Grab*, edited by Alexander Reid Ross. Oakland, CA: AK Press.

Brandeis, Amanda. 2019. "Renewed Hope for Beloved Southeastern San Diego Community Garden." *San Diego Union-Tribune*. https://www.10news.com/news/local-news/renewed-hope-for-beloved--southeastern-san-diego-community-garden.

Brettell, Caroline B., and Faith G. Nibbs. 2010. "Immigrant Suburban Settlement and the 'Threat' to Middle Class Status and Identity: The Case of Farmers Branch, Texas." *International Migration* 49, no. 1: 1–30.

Casey, Edward. 1998. *The Fate of Place: A Philosophical History*. Berkeley: University of California Press.

CEJA (California Environmental Justice Alliance), and California Green Zones. 2021. "National City: The Environmental Health Coalition (EHC)." https://calgreenzones.org/national-city-the-environmental-health-coalition-ehc/.

Chavez, Leo R. 2013. *The Latino Threat: Constructing Immigrants, Citizens, and the Nation*. Stanford, CA: Stanford University Press.

CHIP (Community Health Improvement Partners). 2018. "National City." https://www.sdchip.org/initiatives/resident-leadership-academy/rla-background/national-city/.

Clapp, Jennifer, and S. Ryan Isakson. 2018. *Speculative Harvests: Financialization, Food, and Agriculture*. Nova Scotia, Canada: Fernwood.

Cockrall-King, Jennifer. 2012. *Food and the City: Urban Agriculture and the New Food Revolution*. Amherst, NY: Prometheus Books.

Cutts, Bethany B., Jonathan K. London, Shaina Meiners, Kirsten Schwarz, and Mary L. Cadenasso. 2017. "Moving Dirt: Soil, Lead, and the Dynamic Spatial Politics of Urban Gardening." *Local Environment* 22, no. 8: 998–1018.

Dawson, Julie C., and Alfonso Morales. 2016. *Cities of Farmers: Urban Agricultural Practices and Processes*. Iowa City: University of Iowa Press.

Eizenberg, Efrat. 2012. "Actually Existing Commons: Three Moments of Space of Community Gardens in New York City." *Antipode* 44, no. 3: 764–782.

Fainstein, Susan S. 2001. "Competitiveness, Cohesion, and Governance: Their Implications for Social Justice." *International Journal of Urban and Regional Research* 25, no. 4: 884–888.

Fairbairn, Madeleine. 2021. *Fields of Gold: Financing the Global Land Rush*. Ithaca, NY: Cornell University Press.

Gavidor, Leorah. 2018. "Community Garden to Grow in National City." *San Diego Reader*, January 6.

Gilmore, Ruth Wilson. 2017. "Abolition Geography and the Problem of Innocence." In *Futures of Black Radicalism*, edited by Gaye Theresa Johnson and Alex Lubin. Berkeley: University of California Press.

Hackworth, Jason. 2007. *The Neoliberal City: Governance, Ideology, and Development in American Urbanism*. Ithaca, NY: Cornell University Press.

Hammelman, Colleen. 2022. *Greening Cities by Growing Food: A Political Ecology Analysis of Urban Agriculture in the Americas*. Switzerland: Springer.

Harvey, David. 1985. *The Urbanization of Capital: Studies in the History and Theory of Capitalist Urbanization*. Oxford, UK: Basil Blackwell.

Harvey, David. 1996. *Justice, Nature, and the Geography of Difference*. Oxford, UK: Blackwell.

Harvey, David. 2008. "The Right to the City." *New Left Review* 53: 23–40.

Hernández, David. 2018a. "National City tries to address concerns over community garden." *San Diego Union-Tribune*. https://www.sandiegouniontribune.com/communities/south-county/sd-se-mundo-gardens-20180606-story.html.

Hernández, David. 2018b. "National City to move forward with plans for community gardens." *San Diego Union-Tribune*. https://www.sandiegouniontribune.com/communities/south-county/sd-se-mundo-gardens-20180709-story.html.

Holt-Giménez, Eric, and Justine M. Williams. 2017. "Together Toward Land Justice." In *Land Justice: Re-imagining Land, Food, and the Commons in the United States*, edited by Justine M. Williams and Eric Holt-Giménez. Oakland, CA: Food First Books.

Jewell, Jennifer. 2019. "Robin Wall Kimmerer on Gardening and Citizenship." *Cultivating Place: Natural History and Our Gardens* podcast. https://www.cultivatingplace.com/post/2019/07/04/robin-wall-kimmerer-on-gardening-and-citizenship.

LaDuke, Winona. 1999. *All Our Relations: Native Struggles for Land and Life*. Cambridge, MA: South End Press.

Lefebvre, Henri. 1991 [1974]. *The Production of Space*. Translated by Donald Nicholson-Smith. Oxford, UK: Basil Blackwell.

Lefebvre, Henri. 1996. *Writings on Cities*. Translated by Eleonore Kofman and Elizabeth Lebas. Oxford, UK: Blackwell.

Lefebvre, Henri. 2003. *The Urban Revolution*. Translated by Roberto Bononno. Minneapolis: University of Minnesota Press.

Linebaugh, Peter. 2014. *Stop, Thief! The Commons, Enclosures, and Resistance*. Oakland, CA: PM Press.

Little, Joe. 2018. "Soil Test Shows Toxins Present in National City Community Garden." NBC 7 San Diego. https://www.nbcsandiego.com/news/local/national-city-soil-test/170672/.

McClintock, Nathan. 2014. "Radical, Reformist, and Garden-Variety Neoliberal: Coming to Terms with Urban Agriculture's Contradictions." *Local Environment* 19, no. 2: 147–171.

McClintock, Nathan. 2018. "Urban Agriculture, Racial Capitalism, and Resistance in the Settler-Colonial City." *Geography Compass* 12, no. 6: 1–16.

McClintock, Nathan. 2021. "*Nullius* No More? Valorising Vacancy Through Urban Agriculture in the Settler-Colonial 'Green City.'" In *The New Urban Ruins: Vacancy, Urban Politics, and International Experiments in the Post-Crisis City*, edited by Cian O'Callaghan and Cesare Di Feliciantonio. Bristol, UK: Policy Press.

Myers, Justin Sean. 2023. *Growing Gardens, Building Power: Food Justice and Urban Agriculture in Brooklyn*. New Brunswick, NJ: Rutgers University Press.

Premat, Adriana. 2012. *Sowing Change: The Making of Havana's Urban Agriculture*. Nashville, TN: Vanderbilt University Press.

Pinson, Gilles, and Christelle Morel Journal. 2016. "The Neoliberal City: Theory, Evidence, Debates." *Territory, Politics, Governance* 4, no. 2: 137–153.

Purcell, Mark, and Shannon K. Tyman. 2019. "Cultivating Food as a Right to the City." In *Urban Gardening as Politics*, edited by Chiara Tornaghi and Chiara Certomà. London, UK: Routledge.

Raja, Samin, Marcia Caton Campbell, Alexandra Judelsohn, Branden Born, and Alfonso Morales. 2024. *Planning for Equitable Urban Agriculture in the United States: Future Directions for a New Ethic in City Building*. Cham, Switzerland: Springer.

Ross, Alexander Reid. 2014. "Introduction: The Global Land Grab." In *Grabbing Back: Essays Against the Global Land Grab*, edited by Alexander Reid Ross. Oakland, CA: AK Press.

Rosset, Peter, Raj Patel, and Michael Courville. 2006. *Promised Land: Competing Visions of Agrarian Reform*. Oakland, CA: Food First Books.

Shiva, Vandana. 2014. "Land Wars and the Great Land Grab." In *Grabbing Back: Essays Against the Global Land Grab*, edited by Alexander Reid Ross. Oakland, CA: AK Press.

Shiva, Vandana. 2020. *Reclaiming the Commons: Biodiversity, Indigenous Knowledge, and the Rights of Mother Earth*. Santa Fe, NM: Synergetic Press.

White, Monica M. 2011a. "D-Town Farm: African American Resistance to Food Insecurity and the Transformation of Detroit." *Environmental Practice* 13, no. 4: 406–417.

White, Monica M. 2011b. "Sisters of the Soil: Urban Gardening as Resistance in Detroit." *Race/Ethnicity: Multidisciplinary Global Contexts* 5, no. 1: 13–28.

White, Monica M. 2019. *Freedom Farmers: Agricultural Resistance and the Black Freedom Movement*. Chapel Hill: University of North Carolina Press.

9 Undoing Coloniality Through Nutrition Narratives

Lucy Aphramor

Introduction: Interrogating Public Health Nutrition

Many food justice activists rely personally on the dictates of public health nutrition (PHN) and work to make the foods or diets it deems healthiest more spatially and economically accessible to communities harmed by racism, poverty, and settler colonialism. By PHN, I mean the dominant nutrient-centric and BMI-based philosophy for understanding links between food and human well-being and disease. Activists rely on PHN claims to argue, for example, that food justice will decrease diet-related health disparities (Guthman 2011; Hayes-Conroy and Hayes-Conroy 2013). This outcome is framed as inextricable from the interrelated ambition of reducing rates of fatness, a goal for several authors in this book.

Much of the literature on nutrition and colonialism explores how colonization forced Indigenous peoples away from their traditional foodways and diets as land theft, poverty, soul wounds, cultural repression, and epistemicide led to diets of nontraditional foods and severed food ways (see chapter 3). This is important work. I suggest that its liberatory potential can be augmented by uprooting the deep colonial logic of PHN and cultivating new narratives around food and flourishing. For example, PHN narratives understand the link between (1) high rates of diabetes among Indigenous peoples and (2) the extermination of buffalo through (3) altered nutrient profiles. Here, in PHN's extractivist market economy, buffalo are constructed as meat, and meat is equivalent to a source of nutrients. An alternative expanded, interconnected framing respatializes food and flourishing across buffalo extermination, land theft, poverty, soul wounds, cultural repression, and epistemicide, which all become relevant to high rates of diabetes above and beyond (albeit always inextricable with) any impact on nutrient profile. By separating the inseparable, PHN's imaginaries of disconnect thwart deep change and support the status quo.

I am a white-bodied, queer, Anglo radical dietitian with class and thinness privilege: PHN is the paradigm my professional training taught me to uphold and one that I began to question when people's stories didn't match textbook explanations. In this chapter I illustrate how a conversation around food and well-being can be steered to help people expose and explore some of the hidden assumptions that shape personal and collective affect, ideas, and decisions. My aim is to reveal how PHN mobilizes oppressive ideologies and point to alternative framings that support personal and collective healing and learning enmeshed with reworlding. A key argument is that the deep logic of PHN is a colonial logic and, as such, normalizes and nurtures white supremacy. To help make these links clear, I have capitalized words, like Truth, Rational, Progress, Modernity, Extractivism, Purity, and Perfectionism, that refer to tropes of a colonial logic (Escobar 2008) or that name the hallmarks of white supremacy (Okun 2021).

Using Kitchen Table Pedagogy to Reveal White Supremacy in PHN

Picture a blowzy and buzzy summer day in England. A small group of food justice advocates is sitting along a sickle-shaped curve of straw bales facing a selection of milk containers: a refillable glass bottle for full-cream milk from a local organic herd, a plastic bottle labeled organic semiskimmed milk, nonorganic oat milk from UK oats, organic oat milk from overseas oats grown by a farming cooperative, and fortified organic soya milk.

As a radical dietitian who leads "kitchen-table pedagogy" workshops, I've opened the conversation by inviting people to discuss "Which is the healthiest milk?"

There is an energy of animated uncertainty. In feedback, the first contribution is from a middle-aged man with raised cholesterol levels. In health care he was told this was a significant risk factor for heart disease and was advised to swap to "low-fat" milk, as PHN. He points out that the change seemed unwarranted, as it ignored the fact that he drinks very little milk and his overall diet is low in total fat and saturated fat. So, the generic PHN advice was inappropriate, oversimplified, and erased the particularities of his experience. Even so, PHN gets under his skin and he is trapped in a vexing and taxing conundrum. While he noticed and was perplexed by the disjuncture between his own circumstances and the casually dispensed advice, he nevertheless feels compelled to switch to low-fat milk. But it's not simple: he is thoroughly confused about which milk to reach for, swapping between various plant milks and skimmed dairy milk. It is emotionally and cognitively draining trying to figure things out from so much conflicting information. The doubt and ongoing decision-making take a toll, heightened by fear and shaped by the internalized injunction to keep trying to "get it right"—as if there is a "right" that can be finally arrived at. For our protagonist, an activist, the friction around choosing milk is painfully amplified by recent bereavement; his mother died from heart disease. This tension is compounded by high rates of heart disease among his relatives and at the temple. The speaker recognizes PNH tactics of demonizing food, fat-shaming, and reifying scientific racism in "ethnic" BMI categories as oppressive and abhorrent, but despite being suffused with the critical literature, he's still struggling to interrogate PHN in ways that feel politically useful and that offer different triangulation points for mapping "health" that support personal decision-making.

Why Isn't PHN More Widely Critiqued in Activist Spaces?

PHN is surrounded by an aura of credibility that protects it from due scrutiny. Its claim to be the only valid, safe, and effective way of understanding the links between food and human flourishing, collapsed in a PHN framing to "health," is rendered affectively and intellectually believable wherever scientism prevails. Scientism and science are different: science is a process of rigorous observations and analysis using community-agreed criteria and protocols to produce and share knowledge. On the one hand, as a process of inquiry, the term "science" embraces colonial sciences, Indigenous sciences, social sciences, and more. Scientism, on the other hand, is an ideology that supposes science is the only correct way to conduct inquiry and arrive at knowledge and that the knowledge thus generated is neutral, objective, incontestable, and universally relevant data. Moreover, the science Scientism refers to is limited to that rooted in colonial worldviews. Scientism dismisses global majority sciences as unscientific in a tautological maneuver that secures its own ascendancy and contributes to the ongoing attempted destruction of Global South knowledge, values, practices, lands, and people.

Scientism is not the only means through which PHN exerts its hold on the public imaginary. The fear of illness—personally, in beloveds, and in community—with attendant pain, suffering, financial, caring, and other implications, plus well-founded anticipation of judgement, together with ableism, and possible (self) blame and shame are also powerful drivers for seeking knowledge to inform action.

The ideology of scientism provides context for an ideology called healthism. Healthism is a worldview that maintains that health is a personal property, that lifestyle choice is of the utmost significance to a person's health outcomes, and that supposes we are each morally obligated to prioritize the pursuit of our optimal health (Metzl and Kirkland 2010). To expand, lifestyle choice is generally regarded as pertaining to a person's diet, exercise, and smoking and alcohol habits, and the misplaced choice of the word "choice" is all too often an insult. The "lifestyle choice" headline incorporates the idea that people mainly get heart disease because willful recklessness or unknowing ignorance meant they chose not to drink low-fat milk, for instance. Choosing the "best" milk is a Rational act that enacts our allegiance to Progress and Modernity, simultaneously affirming the moral worth of the consumer for heeding lifestyle advice. The story disappears an entire constellation of factors that impinge on what people (can) eat, as well as expunging another constellation of nonfood factors that materially, metabolically, and spiritually impact the genesis of heart disease. The commonplace PHN interaction of telling someone to change the milk they drink without first ascertaining whether this would be safe, possible, or make any difference is structured around healthism's (i.e., a healthist) logic.

Healthism regards the body as a self-contained calorie-burning machine maintained through prudent lifestyle choices. Thinness (idealized BMI) symbolizes a correct eating and activity regime and is equated with health, blamelessness, and moral worth. Fatness, which is to say fat or larger-bodied people, are coded as inescapably unhealthy, at fault, and morally deficient, leading to serious systemic antifat discrimination referred to by fat activists as fatmisia (Rinaldi et al. 2019). I will return to fat activist scholarship.

The overlapping ideologies of scientism and healthism underwrite PHN's narrative so that it explicitly valorizes individualism, thereby obliterating interconnection and rendering reciprocity, accountability, and reparation superfluous to the health conversation.

Reckoning with Coloniality's Enduring Role in Shaping PHN

I have said that scientism is a powerful driver of what counts as relevant, what is regarded as value-free, and what constitutes truth, making it a vector that perpetuates and creates social norms. The nutrition-science timeline of milk is a case in point. A historical read raises the deep question of "why milk?" and reveals a PHN agenda shaped by the actions of the American Dairy Council and nineteenth-century food evangelists, as well as nutrition scientists (DuPuis 2002). For example, milk was branded a "victory food" and made available in schools in the US to deal with surplus milk from the military after World War II (Belluz 2019). The social history of milk is inconsistent with claims that its place in PHN derives from an unbiased, or even transparent, research process.

This history, together with the campaigns of contemporary food activists depicting today's dairy industry as toxic, oppressive, and violent (Mares 2019; Mendelson 2023), and possibly your own reservations about the way nutrition is taught, suggest that food justice entails disobedience to PHN priorities. It also surfaces the power of questions, and positionality, in reinforcing normative assumptions: my asking about the healthiest milk as a dietitian reinforces a belief system that presupposes milk is a valuable and desirable food.

The Interconnected Body: Heart Disease and Racism

Our speaker in the healthiest milk activity is South Asian (a term generally used in the UK to refer to people of Indian, Pakistani, and Bangladeshi descent) and knows from experience that the rate of heart disease among South Asian men in the UK is about double that of the general population.

According to PHN, the racialized gradient in rates of heart disease is explained somewhat by "genetics" and more so by the "modifiable risk factor" of "lifestyle," which is made indivisible from BMI. The response has been decades of work educating groups at high risk to cook meals that more closely align with PHN guidelines: low-salt, low-fat, "good fat," and high fiber, including adapting a community's traditional cultural meals. By attributing the racialized gradient to lifestyle, PHN fosters a narrative where racism has no salient role in racialized heart disease. This modeling amounts to an extraordinary, and strategic, erasure of power/white supremacy, as diet-related diseases are also, very strongly, power-related—that is, related to oppression and privilege—for example, repressing the strong association between racism and hypertension (Krieger 2014) that is not accounted for by diet and activity.

PHN also removes racism from the field of genetics, wiping out any epigenetic or other change (e.g., framings such as ancestral wisdom) derived from events such as the Bengali famine, redlining, enslavement, and other atrocities. Third, the focus on lifestyle obscures how differences in timely access to medical care and medication, sick leave, and treatment bias account for some of the excess (that is, avoidably early) death and the "excess" life (that is, improved vitality) experienced by those most proximal to white supremacy's standard of Man.

Suppressing the embodied consequences of power, PHN animates a neoliberal analysis constructed around capitalism's atomistic body of homo economicus. Thus, PHN's body-as-calorie-burning-machine is also the hermetically sealed body of liberal humanism, immune to and isolated from the ongoing metabolic bombardment of inequity and historical atrocity. Again, PHN normalizes epistemic violence, ignoring the long lineage of Black scholarship on racism and health that was galvanized by W. E. B. Du Bois and excising the biomedical data that build on this early scholarship. Thus, contemporary PHN ignores decades of biomedical research (deemed high-quality within its own parameters) that points to the overwhelming influence of racism and other social-structural pathways and abuse, rather than dietary pathways, in the genesis of heart disease (Kreiger 2014; McCartney et al. 2012; Felitti et al. 1998; Vitaliano et al. 2002).

Similarly, PHN further embeds racist science in its construct of the sedentary, fat, ghee-eating South Asian man, a stereotype that bears down on racialized, vulnerabilized communities strewing the spores of slow violence viscerally, psychologically, and discursively across time. The obdurate nature of scientism thwarts generative doubt, priming us to accept the principles of healthism and think within models that design out racism and classism. This practice forecloses the possibilities of radically different—that is, abolitionist—responses. I hope that exposing some of the ways scientism and healthism enforce a monoculture of sense-making that frustrates a relational, careful, situated, antiracist sense-making supports the "freedom dreaming" highlighted by the authors and editors of this book.

Naming PHN as a colonial project does not alter the need for people to have decent and dignified meals, however they define that, or for food justice activists and nutrition professionals to play a role in ensuring access to decent and dignified meals and addressing dis/ordered eating, food apartheid, and diet- and power-related diseases for health justice. It might not stop us from judiciously invoking PHN in a crisis or advocacy situation. But it does mean that leveraging PHN and its antifatness agenda as a means to liberatory ends is counterproductive.

Reinstating Eaters, Land, and Relationship

A group of people educated to engage with food through PHN will typically turn first to a consideration of nutrient profile when asked about "healthy food." Scientism runs so deep in our collectively conditioned habits that it can temporarily override what we know to be true and throw us out of alignment with our values.

Centering nutrient profile makes sense in coloniality's divisive, individualistic, anthropocentric imaginary, but not when we permit context. Sometimes the "healthiest" food is simply whatever is available when to forfeit it is to go hungry again. Is a food healthy if it can kill someone with an allergy? If the growers have hazardous working conditions? If it is harvested without ceremony?

PHN primes us to assess healthiness and nutrients as equivalent, an inherent, static, knowable property of a deracinated food. It's not that we are absent an eater, but the eater is made invisible because he occupies the unmarked category of Man. Let's meet PHN's default eater: he is white; has regular, reliable access to all the resources required to safely procure, store, prepare, and eat enjoyable meals without undue challenge or repercussion; is nondisabled, allistic, and not trauma-experienced; does not have an eating disorder, gut condition, limited appetite, poor dentition, or cachexia; and his dietary requirements and preferred eating style and foods otherwise match those recommended and represented. Group efforts to assess the "healthiest milk" use his nutrient needs as pole star, as the default benchmark against which to compare everyone's diet. It's no wonder "universal" guidelines are associated with so much suffering.

As I have said, the disconnect emblematic of "healthy food" is rife across PHN. The unremarkable question "Which is the healthiest milk?" makes food a commodity understood outside relationship with an eater, seasons, governance, food system externalities, and so on: a container for nutrients from nowhere. It also coerces the group into capitulating to an underlying knowledge system that presupposes that "healthy milk" is a meaningful construct and that harnesses colonial tropes of Perfectionism and belief in One Right Way in the insistence that there is a final choice to be made of an all-time *the* best milk.

Antifatness and White Supremacy

PHN's centrifugal force of antifatness presents "tackling obesity" as indisputably advantageous to public health, safety, and plain commonsense. Its self-entailed justificatory mindset posits BMI as a reliable, if imperfect, metric for assessing human well-being and a proxy indicator of someone's diet and activity habits.

Reliance on BMI cultivates fat hatred as a pervasive and unexamined presence that shapes our personal and collective affect, emotion, and beliefs, whatever our size, until we interrogate it. Internalized fat hatred can lead to body shame and dis/ordered eating regardless of our size, and exploring this can be important to healing and sense-making. This is different from the work of heath justice. Health justice, and healing out of whiteness, interconnects personal and social healing/transformation, and this requires naming and rejecting structural, state-sanctioned fat hatred and reckoning with thin privilege and its responsibilities. Fat hatred is at the core of public health and endangers fat people by design. Fat hatred determines health care agendas and literally shapes medical equipment (and seating in schools, buses, theaters, etc.). Fat hatred in biomedicine leads to fat people being refused health insurance, being denied treatment, and receiving substandard care, including missed diagnoses. It means fat people are humiliated, given inappropriate medication and unsolicited lifestyle advice, and subjected to avoidable pain, insulted, and harassed. This barrage of harm is held in place by collective failure to recognize fat people's humanity tied in with uncritical regard for the colonial scientific project enacted through scientism and healthism, for instance, so that we readily collude with its claims of credibility and accept an antifat agenda as value-free science.

The oppressive bearing (and adverse health impact) of BMI-PHN across all aspects of fat people's lives—and fat people's ongoing resistance and endurance—has long been documented in fat activist and fat studies scholarship. This work also shows how BMI is a function of coloniality. Sabrina

Strings's 2019 book *Fearing the Black Body* traces the racist roots of BMI, a classification system invented by the white elite as part of a knowledge regime that constructed the dichotomy of whiteness and Blackness—that is, eugenics. Da'Shaun Harrison's book *The Belly of the Beast* (2021) shows that antifatness can only exist as a coherent ideology with anti-Blackness. This work continues a long lineage of fat (and Black) thought leadership, inside and outside the academy (Pausé and Taylor 2021), that speaks to ethics through distributive justice, anticolonial practice, and abolition.

Building on this, BMI-based thinking requires and stabilizes the mechanistic relationally inert body of liberal humanism inured to the detrimental physiological consequences of living with racism, abuse, war, and displacement—that is, the alienated body of capitalism, conceptually divorced from the physical, spiritual, environmental, and relational costs of wage labor and power hierarchies. Modeling health through BMI emerges from and perpetuates the hallmarks and ideologies of white supremacy and coloniality. This includes work changing obesogenic environments, rather than blaming individuals for, what by implication, is a failed body. The benevolent eradication of fat people through altering living spaces remains a political miscalculation. If the goal is to improve life chances and vitality for fat people in particular, and to improve health justice generally, resources in the UK and US for instance would be more usefully invested in fat-liberatory, disability-forward measures around access, food apartheid, politicized diabetes education, tackling racism and far-right ideologies, universal health care, basic income, employment rights, gender inequity, and other manifestations of "bigotogenic" environments in and beyond medicine and nutrition.

Questions for a Liberatory Pedagogy

Nutrition is a consummate entry point for critical dialogue. We are all eaters or we are fed, we all have/are bodies, and there is popular interest in the food-health connection. Our early eating and feeding experiences profoundly shape our orientation to food, body, and community. Experiences of trauma, including those not directly food-related, likewise have a profound and enduring impact on our physiology and sense of self in ways that significantly complicate eating and safety. Hunger, poverty, food insecurity, racism, ableism, indignities, and dislocation from the land scar many people's lives through present-day and intergenerational materialities and spiritual repercussions. Fat hatred is normalized and still exists in social justice spaces. Food and eating are morally charged and often polarizing, inextricable from class, race, colonialism, dis/ability, and identity. These interconnections and intimacies mean that as food justice activists, we may be, and be talking with, people who have lived through decades of food- and fat-related shame, blame, grief, confusion, and fear.

Perhaps paradoxically, food can become a steadying reference point amid a sense of dispossession. Having an eating regime organized around a solid belief system can help people cope in the absence of other alternatives. For example, food will very often be foregrounded when people live with or care for someone with diabetes. Reframing food can scramble someone's meaning-making system and threaten an entire worldview. Regardless of any liberatory potential, this can be unsettling. There is no "just academic"—as in, knowledge devoid of real-life consequence anywhere—but when it comes to food what we say may have immediate intense and enduring impacts on those who hear us.

On this note, the turn to politicized somatics in movement-building offers a timely and generative opportunity to change food justice work for everyone involved, making it more possible for us to "stay with the trouble" as we consider questions such as How do we work from an anticolonial and antiracist approach, given that many of those most harmed by these systems are dealing with their biomedical consequences (Settee and Shukla 2020)? How can food justice activists and abolitionist nutritionists work with individuals who are scared and angry as hell about the prevalence of diabetes in their community and ask us for "the facts" on diet and blood sugar regulation? With people who are made hungry, feel addicted to sugar, or who live with debilitating digestive problems? And on a more personal note, how can thin, privileged, white-bodied dietitians and other nutrition professionals (like me) usefully contribute to exposing, derailing, and replacing white supremacy and colonialism within PHN without reenacting its oppressive practices? And in doing so, might we also address and begin to heal our own body/fat shame and dis/ordered eating and start the process of professional reckoning and repair?

In this chapter, I have said that the healthy food discourse of PHN derails imaginaries that hold gender equity, antiracism, food sovereignty, pluriversalism, and interconnection as integral to food justice. I showed that healthy eating guidelines are written for the invisible elite white man but taught as goals we should all aspire to, meaning everybody else is harmed by the impact of healthism and "organised abandonment" (Gilmore 2022). Specifically, advice to eat a variety of foods, cook from fresh, avoid processed foods, enjoy eating, and more is ableist. It excludes eaters who rely on shelf-stable foods, prefer a limited range of foods, or for whom eating is so fraught it will always be a chore. It is not an intellectually onerous step to understand ableism in PHN; the challenge lies in shaking off the presumptive belief and associated affect that mean we treat it as a foundationally reliable and always benign regime.

My hope is that the thinking I have shared creates spaces of hesitation in PHN that help us refuse white supremacy and underwrite an emancipatory narrative in its place. The work contributes to and is informed by a long lineage of work by BIPOC dietitians (Dennis and Robins 2020) together with scholars and activists from across fat activism and health at every size (Harrison 2021; Pausé and Taylor 2021), food justice (Higgins 2023; Brady et al. 2019; Hernandez and Luzbetak 2023), critical food studies (Guthman 2011; Hayes-Conroy and Hayes-Conroy 2013) and critical dietetics (Brady 2019), eating disorder activism (Lucas 2023), and disability justice (Wong 2020). To conclude, I want to highlight the report "Gender and Food Sovereignty: Women as Active Subjects in the Provision of Food and Nutrition," where authors already engage food and flourishing through the approach I advocate here, one where "nutrition cannot be separated from food, health, the environment and agriculture . . . [is] comprised of identity, love, care, and spirituality, as well as physical, mental and emotional health" and "integrate[s] the transmission of knowledge, languages, ceremonies, dances and prayers, as well as stories and songs related to subsistence practices and traditional foods" (Galdames et al. 2015).

References

Belluz, Julia. 2015. "How we got Duped into Believing Milk is Necessary for Healthy Bones." Vox. April 19. https://www.vox.com/2015/4/19/8447883/milk-health-benefit

Brady, Jennifer. 2019. "Naming Nutrition Injustice: How Might Dietitians Articulate a Socially Just Dietetic Practice?" *Journal of Critical Dietetics* 4 (2): 6–17.

Brady, Jennifer, Jacqui Gingras, and Katie LeBesco. 2019. "Because . . . Obesity: Reframing Blame in Food Studies." In *Feminist Food Studies*, edited by Barbara Parker. Toronto: Canadian Scholars Press.

Dennis, Mary Kate, and Tabitha Robin. 2020. "Healthy on Our Own Terms: Indigenous Wellbeing and the Colonized Food System." *Journal of Critical Dietetics* 5 (1). https://doi.org/10.32920/cd.v5i1.1333.

DuPuis, E. Melanie. 2002. *Nature's Perfect Food*. New York: NYU Press.

Escobar, Arturo. 2008. *Territories of Difference, Place, Movements, Life, Redes*. Durham, NC: Duke University Press.

Felitti, V. J., R. F. Anda, D. Nordenberg, D. F. Williamson, A. M. Spitz V. Edwards, M. P. Koss, and J. S. Marks. 1998. "Relationship of Childhood Abuse and Household Dysfunction to Many of the Leading Causes of Death in Adults. The Adverse Childhood Experiences (ACE) Study." *American Journal of Preventive Medicine*. 14 (4): 245–258.

Galdames Castro, Mafalda, and María Daniela Nuñez Burbano de Lara. 2015. "Gender and Food Sovereignty: Women as Active Subjects in the Provision of Food and Nutrition." *Right to Food and Nutrition Watch* 31 (3): 29–32

Gilmore, Ruth W. 2022. *Abolition Geography*. New York: Verso.

Guthman, Julie. 2011. *Weighing In*. Berkeley: University of California Press.

Harrison, Da'Shaun. 2021. *Belly of the Beast: The Politics of Anti-Fatness as Anti-Blackness*. Berkeley, CA: North Atlantic Books.

Hayes-Conroy, Allison, and Jessica Hayes-Conroy. 2013. *Doing Nutrition Differently*. NY: Routledge.

Hernandez, Becca Chalit, and Austin Luzbetak. 2023. "Fat in Food and Environment Justice: Lessons from Fat Studies Scholarship." *Fat Studies* 12 (1): 55–71.

Higgins, Alanna K. 2023. "Produce Prescription Programs." *Fat Studies* 12 (1): 21–36.

Krieger, Nancy. 2014. "Discrimination and Health Inequities." *International Journal of Health Services* 44 (4): 643–710. https://doi.org/10.2190/HS.44.4.b.

Lucas, Gloria. 2023. "About Nalgona Positivity Pride." https://www.nalgonapositivitypride.com/about-npp.

Mares, Teresa. 2019. *Life on the Other Border*. Berkeley: University of California Press.

McCartney, G., C. Collins, D. Walsh, and G. D. Batty. 2012. "Why the Scots Die Younger: Synthesizing the Evidence." *Public Health* 126 (6): 459–470.

Mendelson, Anne. 2023. *Spoiled: The Myth of Milk as Superfood*. New York: Columbia University Press.

Metzl, Jonathan, and Anna Kirkland. 2010. *Against Health*. New York: NYU Press.

Okun, Tema. 2021. White Supremacy Culture. https://www.whitesupremacyculture.info/characteristics.html.

Pausé, Cat, and Sonya Renee Taylor, eds. 2021. *The Routledge International Handbook of Fat Studies*. London: Routledge.

Rinaldi, Jen, Carla Rice, Crystal Kotow, and Emma Lind 2019. "Mapping the Circulation of Fat Hatred." *Fat Studies* 9 (1): 37–50. https://doi.org/10.1080/21604851.2019.1592949.

Settee, Priscilla, and Shailesh Shukla. 2020. *Indigenous Food Systems*. Toronto: Canadian Scholars.

Strings, Sabrina. 2019. *Fearing the Black Body: The Racial Origins of Fat Phobia*. New York: NYU Press.

Vitaliano, Peter, James M. Scanlan, Jianping Zhang, Margaret V. Savage, Irl B. Hirsch, and Ilene C. Siegler. 2002. "A Path Model of Chronic Stress, the Metabolic Syndrome, and Coronary Heart Disease." *Psychosomatic Medicine* 64 (3): 418–435.

Wong, Alice 2020. *Disability Visibility: First-Person Stories from the Twenty-First Century*. NY: Knopf Doubleday Publishing Group.

III Encounters

10 Food Assistance Justice: The Discourse and Practice of Food Justice in US Food Banks

Alana Haynes Stein

> As we think about ways to effectively address hunger, there have been kind of three movements in the US. There has been the charitable hunger response, and that's what I would kind of call traditional food banking. There's more of a food systems approach where advocates and actors are really working to build a more healthy, local, sustainable food system that provides both access and affordability to nutritious, culturally appropriate foods to folks at every socioeconomic level and all neighborhoods. And then there's a food justice approach, which really sees food as a basic human right and views all issues of equity and social justice as part of the equation that must be present in order to effectively address hunger. So, we are pivoting away from that charitable response and ensuring that our efforts are pointed squarely at that food justice lens.
>
> —Sharon, CEO of a Feeding America food bank, twenty-plus years in food banking

In the opening quote, Sharon shared how the food bank sector has changed since its inception five decades ago. Sharon described how food banks have responded to broader food system movements, first incorporating alternative food movement values of health, localism, sustainability, and dignity, followed by a recent shift toward promoting food and social justice. In this chapter, I trace the growing food justice movement that I encountered in food banks—describing the development, qualities, and potential of the food justice movement among the institutionalized private charity network.

Food banks are key providers of food to low-income populations and hold considerable power in the private food assistance system. Food banks are nonprofit organizations that source and redistribute food to other nonprofit organizations and increasingly directly to people in need. US food banks are distinguished from food pantries by their roles as intermediaries between donors and other food assistance organizations that only distribute food directly to people. Feeding America holds a monopoly on US food banks with its network of more than two hundred member food banks, twenty-one state food bank associations, and sixty thousand partner agencies providing food to sixty million people in 2020 (Feeding America 2022). Feeding America distributed 6.6 billion meals in the US in 2021 through the redistribution of corporate food waste, federal food assistance signups, federal commodity programs, and purchased food (Feeding America 2022).

Beyond their traditional role as distributors of excess food, food banks are taking on missions to end hunger and poverty through expanded programs and political advocacy. Despite being a key food supplier to low-income populations and playing a growing role in distributing the federal social safety net, the private food assistance network is underexamined in the food justice literature. Perhaps this is due to the critiques raised against food banks for being insufficient, inequitable, and undignified distributors of food and enmeshed in the corporate food system (McIntyre et al. 2016; Haynes Stein 2023; Poppendieck 1998; De Souza 2019; Fisher 2017; Lohnes 2021). Historically, food banks have been criticized for providing corporate welfare to large agrifood businesses, and food assistance organizations have been critiqued for providing stigmatized, racialized access to in-kind food resources that detract from people's right to food (Lohnes 2021; Fisher 2017; De Souza 2019; Haynes Stein 2023; Haynes Stein and Brinkley 2023). While most food banks continue to operate in the ways that have been critiqued, this chapter focuses on a growing minority of food bank administrators (about a quarter of those interviewed) leveraging their privileged positions to promote food justice. They spread food justice through their food banks' influence throughout the broader private food assistance system.

Methods

I interviewed sixty-two food bank administrators, primarily top executives, from fifty-six organizations in the food bank sector via Zoom from 2021 to 2022 until I reached saturation. Forty-seven of the organizations were independent or Feeding America food banks, and nine were related organizations in the food bank sector. I oversampled organizations in the West due to connections through food bank associations but still included organizations from every region of the U.S. (6 percent national, 18 percent

Midwest, 5 percent Northeast, 19 percent South, and 52 percent West). I encountered food justice–oriented food banks in each of these regions, although I did not interview enough people in each region to identify regional patterns. Combined, the organizations in my sample managed over $9 billion of resources according to 2021 IRS revenue records. The median organization had an annual revenue of approximately $80 million ($n = 51$). In addition, I collected secondary materials from food bank websites, *Food Bank News*, 2021 tax returns, mainstream news articles on food banking, and web searches about food banks. I performed web searches about topics that arose in interviews to supplement the interviews and to include organizations beyond the interview sample. Pseudonyms are used for confidentiality.

My experiences and positionality as a highly educated white woman shaped my interactions, generally giving me shared commonalities with the administrators I interviewed, whose demographics were not representative of the food-insecure populations they served. Of the sixty-two respondents, only eleven respondents did not complete the demographic questionnaire, and I was able to pull missing education information from ten of their public professional profiles. Respondents tended to be highly educated (92 percent had at least a college degree, and 51 percent had a postgraduate or professional degree), well-compensated (median salary from IRS records = $163,000, $n = 42$), straight (88 percent), and white (82 percent identified as only white). My sample was 51 percent female and 49 percent male, with no one identifying as nonbinary.

The Context for Growing Food Justice in Food Banks

US food banks have shifted from being organizations with hopes of putting themselves out of business to central providers in the public and private social safety nets. Food bank administrators are increasingly cognizant of their power. Scott, a Feeding America food bank CEO, described the realization that his food bank has a larger budget than regional foundations. This shifted the food bank's approach toward more intentionally working to create systemic changes.

This shift from a focus on feeding people in the present to thinking about ending hunger is prominent in the food bank sector as food banks expand their work beyond providing food. More than half (58 percent) of the organizations represented in my interviews had missions focused on ending hunger or poverty. However, organizations took varied approaches to ending hunger. Among the food banks focused on ending hunger, I identified food banks taking both systemic and individualized approaches to the issue of hunger.

Many food bank administrators claimed they are "ending hunger" by taking incongruous, individualized approaches that emphasize ideals of personal responsibility. These individualistic-oriented food banks suggested they were ending hunger through job training, nutrition education, and cooking classes rather than addressing underlying societal inequities (see chapter 9 of this volume for a further account of nutrition education's injustices). Often, administrators described efforts attached to both systemic and individualistic approaches. Individualistic approaches were discussed in 81 percent of interviews, while food, racial, or social justice only came up in 37 percent of interviews, despite oversampling.

While food and social justice were mentioned in more than a third of interviews, I define administrators as food justice–oriented if they described supporting food justice principles by doing at least two of the following: redistributing power to marginalized groups, intentionally setting up partnerships with BIPOC-led organizations, redistributing organizational resources, providing living wages in their organization, and advocating for social justice–oriented public policies outside traditional social services policies. Of the food bank administrators I interviewed, sixteen (or 26 percent) were food justice–oriented, which included administrators from Feeding America affiliates, independent food banks, and every region. Food justice–oriented administrators were less likely than other administrators in the sample to be straight (69 percent vs. 95 percent) and more likely to have a graduate or professional degree (81 percent vs. 40 percent) Both groups were relatively similar by race and gender.

The inequities of the pandemic and the concurrent 2020 Black Lives Matter movement led food banks to increase their attention to equity and social justice. Food justice–oriented food banks generally began or ramped up their food justice and equity work during the pandemic. Food bank administrators emphasized how the pandemic and the 2020 Black Lives Matter movement prompted change. William, a Feeding America food bank CEO, said the Black Lives Matter movement "increased the urgency for the work around equity in a big way" while also providing "new language to talk about the racial injustice and the impact that our societal systems have on people of color and Indigenous people across the country." The COVID-19 pandemic also forced food banks to implement changes. They adapted to public health orders while being inundated with new food and financial resources from government programs and private donors. In response, food banks reexamined their service models, which led to new opportunities to incorporate food justice.

Food Justice–Oriented Food Banking

Food justice–oriented food banks emphasized their critical roles in simultaneously meeting the immediate needs of people experiencing hunger and working to change the systems perpetuating hunger. Putting food justice front and center, Kevin said his role as CEO of a Feeding America food bank is "to enact not only food security but food justice." Lauren took a rights-based approach as CEO of a Feeding America food bank, saying, "Food is a basic right." She described her food bank's use of a racial justice framework to examine its role in perpetuating systemic racism. Lauren's food bank is examining the "deep-rooted historical ties that disenfranchise certain parts of the population, how that plays out with our programming, and whether or not we're contributing or dismantling those . . . types of racial undercurrents that have existed in our society for so long."

As food banks seek to promote food justice, they are changing their missions to reflect their new values rather than their prior focus on supplying food. While 78 percent of my sample still discussed traditional ideas of providing food or alleviating hunger in their missions, 15 percent had missions featuring systemic change or equity. Encountering food justice within mission statements demonstrates how food justice has become a central value within these food banks. For example, Washington-based Northwest Harvest proclaims its new mission is "growing food justice through collective action." As another example of this trend, the National Feeding America Organization changed its mission in 2021 to emphasize creating change to support equitable food access. This replaced its former mission "to feed America's hungry" and engage the public.

In addition to changing missions, justice-oriented food banks are also changing their programs and structures. These changes include reorienting programming to support food justice through diversifying their organizations and networks, redirecting resources toward programs that support a more equitable food system, and advocating for policy changes and social movements to promote social justice.

Shifting Power Relations

Food banks are advancing their new missions by bringing people from diverse communities that have experienced hunger into decision-making roles and intentionally setting up relationships with BIPOC-led organizations. With increased food and financial donations alongside growing national attention to diversity, equity, and inclusion (DEI), food banks have had more donor support for food justice work. Sharon explained that her organization was "deeply invested in our equity journey before the murder of George Floyd" but that the movement "really helps us become bolder and to take more regular, consistent external actions."

Food justice–oriented administrators are working to shift the power dynamics within their food banks by including diverse groups of people who have experienced hunger in decision-making, a practice discussed in 29 percent of interviews. Jennifer, a Feeding America food bank CEO, described her food bank's new program that provides public speaking training to people who have experienced hunger and connects them with media outlets and politicians. She explained how the program has also changed decision-making within the food bank, saying, "They are *clearly* the subject matter experts. We learned way more from them than they ever learn from us." Jennifer described how this new program shifted her food bank's practices. She said the growing national awareness "that those in power of resources also have all the power of the decision-making" pushed the food bank to center the voices of people who have experienced hunger, resulting in devoting more resources to political advocacy.

Food banks are also diversifying the organizations they work with. Historically, food banks primarily distribute food through their partner agency network. These networks are predominately faith-based organizations and have primarily developed based on which organizations approached the food banks, leading to patchy and inequitable networks. More than half of food bank administrators (58 percent) described recent targeted efforts to distribute food to high-poverty areas and communities of color.

Food justice values were apparent in organizations that intentionally formed partnerships with BIPOC-led groups and organizations, a practice described by 26 percent of administrators. Kevin spoke of his food bank's work with local Indigenous groups and his food bank's reflexivity, that they "don't want to replicate some white saviorism and white dominant culture, which nonprofits like the food bank can often be part of that racist charity industrial complex that takes place in the nonprofit sector." Kevin's food bank sought to differentiate itself from traditional food banking through intentional relationship building, centering food justice values, and work focused on building a more just food system.

Investing in a More Equitable Food System

Food justice–oriented food banks are also taking innovative approaches to expand and rethink their programming by redistributing resources. For instance, Kevin detailed how his food bank pushed beyond traditional food banking by supporting community gardens and investing in food entrepreneurs. Other food banks directed new funding toward BIPOC-owned and -led businesses, farms, and nonprofits. Over a quarter of food bank administrators (26 percent) portrayed organizational efforts to redistribute resources to support a more equitable food system.

These programs included purchasing gift cards from BIPOC-owned grocers, purchasing food from BIPOC-owned farms, and providing capacity-building grants to BIPOC-led nonprofit organizations. Jennifer described how her food bank established a new program to redistribute resources more equitably in recognition of the systemic racism and inequities that undergird resource distribution in the charitable food assistance network. Jennifer said, "[The program] is specifically for organizations that are led by and serving communities of color, who are likely not part of the food bank network. And so, this is just recognizing that many of these grassroots organizations, they're serving some of the most acute needs of food insecurity, and that they frequently do not receive the same kind of funding that organizations like [our food bank], a very long, well-established, very much a[n] all-white led organization tends to get more funding.

Andrea, an independent food bank executive director, described her food bank's efforts to change their local food system by building resilience through permaculture farming education, prioritizing relationships with BIPOC groups, and focusing on land back efforts with local Indigenous groups (see chapter 5 of this volume for more on how repatriating land can improve Indigenous food sovereignty and health). These efforts were at the forefront of food banks working to strengthen local food systems rather than relying on the corporate agrifood system that undergirds traditional food banking and perpetuates poverty wages (Lohnes 2021; Fisher 2017; Haynes Stein and Brinkley 2023). Justice-oriented food bank administrators were cognizant of how they could use their power to push back against corporations and politicians who have traditionally used food banks to distance themselves from their role in creating and perpetuating hunger.

Political Advocacy

Although nonprofits are often considered apolitical entities, 98 percent of the administrators I interviewed spoke of dialoguing with elected officials or working to influence government policies, program administration, and funding. Food justice–oriented food banks more frequently highlighted how both hunger and food banks are outputs of public policies and capitalism. Scott said, "Hunger is a function of food insecurity, food insecurity is a function of poverty, poverty is a function of racism, and then just general economic inequity. If people weren't poor, they wouldn't be food insecure. They wouldn't be coming to us. So, ultimately, we see this as a political-economic question." Scott emphasized that hunger results from an inequitable system that must be remedied through policy rather than food assistance. Through framing hunger as a political choice, he highlighted the political role food banks must take to end hunger.

Most food banks (91 percent of those interviewed) mentioned involvement in political advocacy to support public and private food assistance programs. Only a minority (16 percent) described expanded legislative agendas that advocated for systemic change, supporting policies that broadly benefit communities marginalized by race and income beyond food assistance and social welfare policies. Administrators described advocating for social justice—oriented policies that broadly affected people in poverty and the food system, such as living wages, the right to food, decriminalizing drug possession, police accountability, changing land use policies in support of urban gardening and farming, Black maternal health, and promoting access to childcare. Beyond policy work, food banks also supported environmental and social justice movements like antipipeline protests and the Black Lives Matter movement.

Within these social justice–oriented policies, food banks are particularly turning toward wages. Inadequate wages came up in more than half of the interviews, and 26 percent of interviewees discussed living wages. This represents a powerful turn, as much of food banks' support historically comes from agribusiness and grocery retailers who profit from low wages.

Five interviewees discussed starting their wage-related advocacy by revising their employment practices to pay living wages. Food banks also press for increased wages in their communities through public forums and public policy initiatives. The Community Food Bank of Southern Arizona sponsored a local ballot initiative for a $15 minimum wage in Tucson (Verwys 2021). Still, these initiatives faced pushback, as evidenced by a municipality outside the policy's jurisdiction that pulled funding from the Community Food Bank of Southern Arizona due to the food bank's ballot initiative contribution (Verwys 2021).

Bringing Power to the Table

As food bank administrators incorporated food justice into their work, they highlighted the immense power food banks bring to the table based on their institutionalization, economic resources, and social/political ties. Food banks are central distributors of multiple federal assistance programs, and the US food bank system is a multi-billion-dollar industry with a massive base of donors, volunteers, and clients. Beyond their organizations' capital, food bank administrators also tended to be privileged in their racial, financial, and educational positionality. Most interviewees had annual incomes over $100,000 and were white and well educated.

Kim, a Feeding America food bank CEO, described her food bank's political advocacy work as a strategic initiative to build political power in support of systemically marginalized populations. Kim outlined how hunger is a systemic problem tied to racial and class oppression. She said, "To pull those barriers down, you need political power." Kim described her food bank's strategy of symbiotically using in-kind food assistance to build an organized political movement. She emphasized how her food bank builds social movement relationships and community through food.

Justin, a Feeding America food bank CEO, described how he used his position as a public figure invited to a large university's food insecurity summit to critique detractions from the root causes of hunger. He brought a critical perspective, pushing other leaders to consider the economic inequality that causes hunger. Justin described his interactions with the university's leadership and how he confronted the university for perpetuating the root causes of hunger through low wages and privatization.

> I was sitting with the Chair of the Board of [the university], and I said, "You're not gonna like what I say for part of this." He goes, "What?" And I went, "You got to quit privatizing the bottom of your workforce. You're creating the problem. People, they simply don't have enough resources, and you made it worse because now they lost tuition reimbursement," which was a component for everybody that worked at [the university.] And I said the same thing to [a large insurance firm], to some banks, "At the end of the day, your wages are too low at the bottom."

Justin tied the issue of hunger to capitalism and inequality in the US, calling out people in power responsible for these changes. Justin used his position of power to contradict the assumptions that the charitable food system could solve the problem and to push people in power to raise wages and benefits. Similarly, Scott talked about how the food bank could better wield its power to promote systemic change. He highlighted how the food bank has 85 thousand volunteers a year that they could influence to effect systemic change. Other food banks also described their efforts at leveraging their positions in the community to create systemic change. As Jennifer previously discussed, food banks' organizational power has been accrued through the racial capital of their organizations, which some administrators are beginning to focus on equitably redistributing.

Hegemonic Food Banking

This chapter focuses on the incorporation of food justice within food banks, yet food justice–oriented food banks are still in the minority. Despite oversampling food justice organizations, only about a quarter of administrators were food justice–oriented. Most food banks still reflect the many critiques levied against food banks as they provide tax benefits and PR to corporations while providing inadequate, stigmatized food through inequitable means (Lohnes 2021; De Souza 2019; Fisher 2017; McIntyre et al. 2016; Poppendieck 1998). While I spoke to food bank administrators who were working to create a more equitable society, most food bank administrators I interviewed (79 percent) spoke about highly individualistic programs that placed blame on their clients and pushed clients to pull themselves up by their bootstraps through cooking classes, nutrition education, budgeting workshops, and job training programs for low-wage food system work (see chapter 17 of this volume for more on food system worker exploitation). Most food justice–oriented organizations (87 percent) also operated at least one of these individualistic programs, showing how food justice is still just one of several approaches to addressing hunger that might be encountered at food banks. In the larger population of food bank administrators, I spoke to administrators who espoused highly stigmatizing and discriminatory practices, requiring clients to jump through hoops to demonstrate need and pushing religion along with food assistance. While I did encounter a sizable minority of food banks advancing food justice, most food bank administrators (74 percent of those interviewed) are not yet focusing substantial attention on these efforts.

Food justice–oriented administrators are also still wrestling with how to increasingly focus on food justice while still addressing the immediate need for food that their organizations have historically filled. Sharon described how her firsthand food insecurity experience informed her approach as a food bank CEO. She said, "We can't abandon our moral obligation to feed people experiencing hunger today, to *only* focus on the ending hunger work. I believe we need to do both." Sharon called upon the moral imperative to continue feeding people now. However, she advocated for an approach that places far more emphasis on systemic change as opposed to the charity model of traditional food banking she critiqued in the opening vignette. In the present highly polarized political climate, the future of food banking is uncertain. Food banks may continue to grow their food justice work, or they may continue to support the inequitable food industry that their programs have historically subsidized.

Conclusion

Notably, food justice–oriented food bank administrators do not believe traditional food banking is the solution to ending hunger or creating a just food system. They view themselves as shifting their organizational approaches to participate in the broader food justice movement—changing the unsustainable and inequitable global agrifood system. Food banks have considerable power and privilege, which affords them the potential to play an outsized role in shifting the status quo. They have taken important steps to shift power and decision-making to those who have experienced hunger, redistribute resources to BIPOC-led businesses and organizations, and advocate for social justice policies. Still, these food banks are confronting the legacies of their organizations that are entrenched in the charitable model and individualized approaches to addressing poverty. Food banks' incorporation of food justice has also remained relatively limited in scope, mirroring the early food justice movement (see introduction of this volume).

While some food banks are beginning to focus on inequities in populations that disproportionately experience hunger, most of these conversations focused on race/ethnicity and class. Other populations that disproportionately experience hunger should not be overlooked. LGBTQ+ communities were rarely mentioned despite experiencing disproportionately high rates of food insecurity (see chapter 2 of this volume). Several food bank administrators also described the increased vulnerability of farmworkers and how their organizations designed programs to reach farmworkers. Despite this targeted outreach with food assistance, systemic changes to farmworker conditions were largely absent from the conversation (see chapter 17 of this volume for further discussion of the importance of improving labor conditions). Both within and outside food banks, the food justice movement should pay more attention to the diverse populations most impacted by food injustice.

Food banks are still largely a part of the charitable industrial complex. Proponents should ensure that politicians and corporations do not use food banks to uphold exploitation and detract from a dignified and comprehensive social safety net. The traditional food bank model is not a model for food justice. However, with the considerable power food banks wield in US society, my encounters with food justice within the institutionalized US food bank system point to a powerful potential ally for the food justice movement.

References

De Souza, Rebecca. 2019. *Feeding the Other: Whiteness, Privilege, and Neoliberal Stigma in Food Pantries*. Cambridge, MA: MIT Press.

Feeding America. 2022. "2021 Annual Report: Doing More to End Hunger." https://www.feedingamerica.org/sites/default/files/2022-01/FA_2021AnnReport_FULL_012522.pdf.

Fisher, Andrew. 2017. *Big Hunger: The Unholy Alliance Between Corporate America and Anti-Hunger Groups*. Food, Health, and the Environment series. Cambridge, MA: MIT Press.

Haynes Stein, Alana. 2023. "Barriers to Access: The Unencumbered Client in Private Food Assistance." *Social Currents* 10 (3): 207–224. https://doi.org/10.1177/23294965221129572.

Haynes Stein, Alana, and Catherine Brinkley. 2023. "Farm to Food Bank: Exploring the Ties Between Local Food Producers and Charitable Food Assistance." *Rural Sociology*. https://doi.org/10.1111/ruso.12489.

Lohnes, Joshua. 2021. "Regulating Surplus: Charity and the Legal Geographies of Food Waste Enclosure." *Agriculture and Human Values* 38 (2): 351–363. https://doi.org/10.1007/s10460-020-10150-5.

McIntyre, Lynn, Danielle Tougas, Krista Rondeau, and Catherine L. Mah. 2016. "'In'-Sights about Food Banks from a Critical Interpretive Synthesis of the Academic Literature." *Agriculture and Human Values* 33 (4): 843–859. https://doi.org/10.1007/s10460-015-9674-z.

Poppendieck, Janet. 1998. *Sweet Charity? Emergency Food and the End of Entitlement.* New York: Viking.

Verwys, Jamie. 2021. "Sahuarita Cuts Ties with GV Food Bank over Campaign Donation." *Green Valley News & Sahuarita Sun*, October 16, 2021. https://www.gvnews.com/news/sahuarita-cuts-ties-with-gv-food-bank-over-campaign-donation/article_9759a820-2dd0-11ec-bcff-abf945cbda41.html.

11 Street Food Justice

Charlotte Glennie Roberts and Catherine Brinkley

Introduction

Food justice depends on affordability and access. People need healthy, culturally appropriate food options that fit their budget, and they need those food sources within easy reach of home. Policymakers, scholars, and food-focused foundations have mostly envisioned meeting this goal with full-service grocery stores (Pringle 2013; Brinkley et al. 2019). However, efforts to bring grocery stores to communities with low food access have had mixed results, despite the generous tax breaks and grants that are often used to attract a new supermarket (Brinkley et al. 2019).

Amid the vigorous public conversation around food justice, food deserts, and strategies to increase food access, one promising tool—street food vending—has been largely overlooked. Street food vending is the practice of selling food items in open-air spaces including parks, sidewalks, parking spaces, parking lots, or temporarily closed streets. Street food vending can take many forms, from "pop-up" food retail, such as pushcarts and farmers markets, to more stable fixtures in the built environment, such as regular sales from parked food trucks and box trucks (Brinkley et al. 2013). As noted by Allison, Ray, and Rohel (2021), street food vending contrasts with the supermarket model in significant ways: it sustains more livelihoods, contributes less to the concentration of wealth through profits, broadens food access, and cultivates social ties and trust at the neighborhood level. Street food vending facilitates encounters between different kinds of people who might not otherwise interact, and this chapter will show how vendors from marginalized communities are further asserting their visibility and humanity through public campaigns to legalize street food.

This chapter highlights the potential of street food vending to increase food justice, details the struggle of street food vendors to gain legal status for their work, and points to critical ways that researchers and policymakers can facilitate healthy street food vending as a way to improve food access. Street food vendors are organizing for better policy in cities across the US; while the legal barriers and campaigns to address them vary by city, this chapter focuses on Los Angeles as an instructive case study.

Street Food in Scholarship

Evidence suggests that street food vendors can improve public health by bringing affordable produce where it is needed most. With low startup costs and the physical mobility of their operations, street food vendors can succeed in neighborhoods that don't have the purchasing power to support restaurants or full-service grocery stores (Yasmeen 2001; Francis and Brinkley 2020). Street vending contributes to urban food security, as it is often the only affordable food source close to the homes or workplaces of poor and working-class urban residents (Hayden 2021). Compared to the closest supermarkets, curbside vendors that sell produce offer similar variety at lower prices (Brinkley et al. 2013). Street food vendors contribute to lower produce prices throughout the food environment by adding competition, and they also increase the visibility of fruits and vegetables, encouraging healthy food choices (Li et al. 2014). Further studies support the finding that street food vending encourages healthy eating habits: Zepeda et al. (2014) found that street food customers eat more fresh produce on average than noncustomers, and Breck et al. (2015) found that street food vendors are the most likely produce source for Supplemental Nutrition Assistance Program (SNAP) recipients, who spend an average of $3.86 more on produce than customers paying with cash. Such findings vary by geography and over time, but they hold promise.

Street food vending also promotes food justice through economic development. Low startup costs and overhead mean that street food vending offers a small business opportunity for micro-entrepreneurs with little access to capital. Women and immigrants are disproportionately represented in the street food vending economy (Yasmeen 2001; Calderon 2015). Since street food vending does not necessitate a specific education level or language fluency, it has fewer barriers to employment than other entry-level jobs, and when vendors are self-employed, they can set their own hours—an important flexibility for caregivers (Bennett et al. 2021). Street food vending thereby contributes to food justice as an income-generating opportunity for those with limited job options. Furthermore, street food vending benefits the local economy through the vendors' own spending on supplies and ingredients, which are often purchased from other local businesses (Liu et al. 2015).

Despite these health and economic benefits, much of the public health literature on street food focuses instead on food safety concerns (Francis and Brinkley 2020). Restaurants, schools, and other food-serving institutions have not faced the same scrutiny in the public health literature (Ababio and Lovatt 2015), despite the risks being comparable between mobile and fixed food establishments (Faw and Tuttle 2014; Donkor et al. 2009). In sum, the public health literature's emphasis on concerns about the safety of street food over and above the benefits has helped advance a narrative that street food is particularly unsafe and needs to be heavily regulated or outright banned—a narrative that coheres with long-standing racist tropes (Elliot 2015) to justify the "differential treatment" of immigrant and working-class street food vendors (Agyeman, Matthews, and Sobel 2017; see also Martin 2014 and Lemon 2019).

Restaurant industry representatives, business associations, and local politicians have most frequently invoked public health and food safety concerns when advocating for greater regulations on street food vending. In many jurisdictions across California and beyond, onerous regulations have been imposed on street food vendors, frequently under the guise of public health considerations (Francis and Brinkley 2020). However, given the limited empirical data supporting the notion of street food as especially unsafe, and the dubious connection between many specific restrictions and food safety, the primary motivation for such policies appears to be the perceived threat that mobile food vendors represent to brick-and-mortar food businesses.

To date, little scholarship has critically examined street food vending policies and their relationship to the public health literature on street food. The authors of this chapter, two white mothers with long-standing interests in food justice, seek to fill this gap as part of our commitment to building healthy and accessible food systems. We did not participate directly in the Los Angeles Street Vendor Campaign, nor did its members have input in the development of our arguments, but we did interact with them in our roles as researchers and sought feedback on this chapter before publication. Our discussion draws from spatial quantitative analysis undertaken by Francis and Brinkley (2020), interviews with members of the Los Angeles Street Vendor Campaign, and archival analysis of public hearings, newspaper articles, reports, and policy documents to build out a case study focused on the community development efforts that have underpinned the legalization of street food in Los Angeles.

In the following section, we recount the efforts of street food vendors in California to resist the criminalization of street food and put forth an alternative vision for healthy food policy. This movement is instructive both for food justice scholars and activists who may have given little attention to street food vending thus far and for public health researchers who may not have considered the health and economic benefits of mobile food vendors operating in areas with low food access.

The Street Food Justice Movement in California

Street food vending has been an integral part of Los Angeles culture since at least the 1870s, despite hostility from brick-and-mortar food businesses and racialized criminalization of vendors occurring over much of this history (Elliott 2015). Selling food and other goods on the city's sidewalks, parks, and beaches has been largely illegal for decades, yet an estimated 10,000 street food vendors continue to work across the city (Liu et al. 2015). Periodically, police have increased enforcement efforts and issued fines or confiscated these sidewalk vendors' carts and supplies, with the threat of further legal action being especially frightening for the many street vendors without legal status (Abello 2022). In 2008, police initiated one such crackdown and ultimately shut down a night market in the Boyle Heights neighborhood; facing the loss of this thriving street food hub, vendors and advocates with the East Los Angeles Community Corporation began to organize for legalization and permitting (Abello 2022).

The vendors and their allies, including food justice scholars in the local research community, held community forums and met with city councilmembers, and in 2012 they formed the Los Angeles Street Vendor Campaign (LASVC). From the first introduction of a motion to permit street vendors in 2013, it took five years for the Los Angeles City Council to finally establish a pathway for vendors to operate legally.

Facing continued challenges at the local level, advocates with the LASVC organized with vendors from across California to change policy at the state level. As Los Angeles officials were sorting out the details of their new permitting process, the state passed SB 946 (2018), the Safe Sidewalk Vending Act. This law prevents cities and counties from imposing restrictions on street vending that are unrelated to public health or safety.

Our research on street food policies across California showed that after the passage of SB 946, most cities and counties throughout the state were out of compliance with the law (Francis and Brinkley 2020). Going beyond restrictions related to public health and safety, 65 percent of cities limited the time vendors could stop in any one location, forcing them to move around with often unrealistic and prohibitive frequency; 54 percent of cities restricted vending activities by location, such as banning street food vendors from parks; and 43 percent of cities imposed labor requirements unrelated to public health and safety. For example, sixteen cities required a social security number to be listed on vending permit applications, a policy that is expressly prohibited by SB 946 because of its disproportionate impact on immigrants. Furthermore, sixty-eight cities required criminal background checks and fingerprinting for all vending permit applicants, requirements that are not imposed on restaurant owners or other fixed retailers and present an added barrier for undocumented and formerly incarcerated people seeking to earn a living through street vending. Such policies are testament to the long-standing resistance that street food vendors have faced from brick-and-mortar businesses seeking to reduce competition, which have stoked a narrative about street food as unsafe alongside racist dog whistles about "protecting neighborhood character" (Elliot 2015; Abello 2022).

While organizing to build the political power that achieved street vending legalization in both Los Angeles and the State of California, the LASVC also worked to change the conversation around street food vending by advancing a counternarrative about the social, cultural, and economic impacts of street food. Partnering with researchers at the Economic Roundtable and the UCLA School of Law Community Economic Development Clinic, LASVC and its coalition members produced reports on quantifying the contributions made and challenges faced by street food vendors in Los Angeles (Liu et al. 2015; Bennett et al. 2021). The first report, "Sidewalk Stimulus," explicitly challenged the notion that street vendors are a threat to public safety or are unfair competitors drawing business away from brick-and-mortar establishments. According to the report, street food vendors generate $1.60 in direct and indirect economic impacts for every dollar earned. Furthermore, their geographic analysis found benefits rather than detriments to nearby fixed retail establishments: "Brick and mortar businesses benefit from the presence of street vendors. Few in Los Angeles shared the same city block as a sidewalk seller, but those that were collocated with vendors were less likely to close their business over the course of the recession. Retail stores and restaurants near vendors were also more likely to expand employment from 2007 to 2011. Not only do street vendors attract more foot traffic to brick and mortar shops, but their presence is linked with safer streets." (Liu et al. 2015, 20). The second report, "Unfinished Business," focused on policy barriers remaining after the passage of SB 946 that left most street food vendors unable to obtain permits for their operations. Before detailing the barriers to permitting, the report summarizes individual and community-level benefits that street food provides, emphasizing vendors' multidimensional significance: "Sidewalk vendors provide a way to celebrate and express culture in historically disinvested communities. In these instances and many others, sidewalk food vendors are far more than a cultural reference or a tourist's novelty in one of the nation's largest and most diverse cities: they are engines of economic productivity and pillars of their communities." (Bennett et al. 2021, 5) In this report, as in "Sidewalk Stimulus," the authors framed the value of street food as a vital component of Los Angeles culture, a lifeline to healthy food access for low-income residents whose neighborhoods lacked grocery stores, and a vibrant part of the local economy. Profiles in these reports, advocates' public statements, and social media content produced by the LASVC have consistently sought to humanize street food vendors, draw attention to their contributions to public life, and highlight the consequences of their criminalization through poor policy (see figure 11.1).

Through its organizing, the LASVC has also put forth an alternative vision and policy platform that supports street food as a legitimate component of the food system, ensures the safety of both food and the vendors who provide it, and increases the accessibility of street food. For example, as the Los Angeles City Council was determining how to structure rules and regulations for street vending, the LASVC advocated for a system including permits and supported rules to ensure that vendors prepared food safely and cleaned up the sites where they operated before leaving. The LASVC's Facebook page

Figure 11.1
The Los Angeles Street Vendors Coalition rallies on the steps of city hall. The banner reads, "We also contribute to the economy," and the popsicles say "respect," "justice," and "solidarity." Photo by Rudy Espinoza, executive director of Inclusive Action for the City.

lays out their goal for city policy as one that will "permit food vending on sidewalks, permit non-food vending on sidewalks, incentivize healthy food vending, create partnerships with small businesses to cultivate vibrant commerce, foster safer streets by putting more 'eyes on the streets,' and create jobs for low-income, unemployed workers" (LASVC 2023). Public hearing testimony from street food vendors and their allies framed legalization of street food as a matter of economic inclusivity for hardworking microentrepreneurs. Vendor and single mother Isabel Rodriguez testified about an incident where she had been ticketed and handcuffed in front of her son, explaining, "We sell out of need, and if we stop selling, our families will suffer. Work is not a crime. We want to live in peace" (LA City Council joint committee meeting, November 8, 2017). This framing challenged the prevailing dichotomy between "vendors" and "businesses" in testimony from brick-and-mortar business advocates and some city councilmembers, recasting vendors as members of the business community who have faced unjust exclusion and repression due in part to race, gender, and class inequalities. Throughout their campaign, LA's street food vendors have made it clear that they simply want to be able to work legally within a system that promotes health, minimizes conflict, and ensures safety.

Testimony from Los Angeles city council meetings also highlights the extent to which food safety concerns have permeated the public discourse around street food vending. When council members voted on November 28, 2018 to pass the sidewalk vending policy proposal, two of seven councilmembers speaking before the vote explicitly invoked public health and food safety, while three others referenced "safety" in general as a justification for the policy. Furthermore, public comment from brick-and-mortar business advocates tended to invoke food safety when calling for policy elements opposed by vendors. Yet while comments invoked food safety concerns, policy solutions often focused on giving propertied business owners power over street vendors. For example, at a city council committee meeting deliberating on details of the street vending policy, a representative from the Historic Core Business Improvement District asked, "How can property owners and businesses ultimately opt in or out, essentially weigh in, as they would need to anyway to ensure proper public health requirements are met?" (LA City Council joint committee meeting, November 8, 2017). The logical connection between brick-and-mortar control over street vendor access and how it would improve public safety is threadbare. Nevertheless, in 2018, shortly before the ordinance was passed, a representative with the Central City Association affirmed the value of restrictions on vending around businesses and entertainment venues as "necessary to protect the health and safety of all who frequent these destinations" (LA City Council joint committee meeting, November 27, 2018). In December 2019, a representative from the SoCal Restaurant Association gave an impassioned speech about the need to distinguish between pushcart vendors and "mobile restaurants" with tables and chairs (which were never legalized under the city policy), saying that "if somebody eats tainted food, they may die, have a fever, go to the hospital and not know why they died; you will be directly responsible because you failed to listen . . . the vendor act has to be amended" (City Council meeting, December 4, 2019). Though such arguments do not hold up under close scrutiny, they demonstrate how public health concerns have repeatedly been invoked as brick-and-mortar businesses have sought to reduce competition from street vendors. Advocates for fixed food retailers and property owners were unable to construct reasonable claims about public health concerns in part because the LA Street Vendor Campaign had explicitly embraced the need for food safety regulations from the outset.

A decade on, this campaign has achieved significant victories in legitimizing street food vending in Los Angeles and statewide; however, the gains have been incremental and required sustained organizing to overcome a series of policy challenges. LA's street vendor permitting program was just getting up and running when the COVID-19 pandemic hit. On March 17, 2020, one day after Governor Newsom issued guidance for restaurants to stop dine-in service, the LA City Council took up a motion to temporarily ban all street vending. Because this move would have been in violation of SB 946, the council passed an amended version that allowed fully permitted vendors to continue operating—however, less than 0.5% of the city's street food vendors had received the necessary permits at that point. Thus, almost all of the city's street food vendors were subject to a new wave of police crackdowns at a time when economic precarity and food insecurity were skyrocketing.

As pandemic restrictions eased and city services resumed normal functioning, the pace of street food vendors obtaining their permits in Los Angeles remained painfully slow. As detailed in the "Unfinished Business" report, vendors found that getting the health department's sign-off was nearly impossible given stipulations in the California Retail Food Code, which did not have provisions for sidewalk pushcarts. Instead, the guidelines for street food vending were written with food trucks in mind. Sidewalk vendors who attempted to meet the requirements found that an up-to-code pushcart would cost at least $5,000, prohibitively expensive for many vendors, and would weigh 1,200–1,600 pounds, virtually impossible for most people to push (Bennett et al. 2021).

With the release of the "Unfinished Business" report, the LASVC undertook a new effort to amend policy at the Department of Public Health and the state and city levels. Throughout the report and their wider efforts, the LASVC was clear in seeking policy that would preserve evidence-based public health standards while making permitted sidewalk food vending a viable business. They also emphasized the ways in which their proposed policy changes would create new economic opportunities for nonvendors, such as builders who could design and mass-produce carts to meet local public health specifications and restaurants and churches that could rent kitchen space for street vendors to prepare their foods (Abello 2022). In response to street vendors' organizing efforts, the state passed SB 972 in September 2022. The bill amends the California Food Retail Code to create a new category for sidewalk food vendors, "compact mobile food operations," establishing appropriate equipment standards for human-powered pushcarts and allowing a wider range of facilities to be approved as sites for vendors' food preparation and storage (Public Counsel 2022). Like SB 946, the new law also works to decriminalize street food vending by reducing code violations from a criminal to an administrative offense. The policy clearly reflects the concerns and priorities of sidewalk food vendors that the LASVC had documented and amplified through its organizing activities.

Seeking a More Just Future Through Street Food

The case of the Los Angeles Street Vendor Campaign shows a promising example of how people are organizing to build a more just food system. Street food vending is largely entrepreneurial; workers are not pursuing a unionization strategy like those described in chapter 17, but they are nonetheless contributing to the wider resurgence in campaigning for labor rights among food-system workers and their allies. The practice of street food vending, and in particular, vendor-led campaigns to legalize street food, shows another path by which marginalized workers in the food system are pursuing "labor sovereignty" as described in chapter 18. However, across

California and beyond, significant barriers still prevent street food vendors from operating legally and maximizing their ability to offer healthy, affordable food where it is needed most.

In Los Angeles, community organizations have had to shoulder much of the work of educating people about the policy changes. Not only are advocates communicating with street vendors and helping them navigate the still-complex permitting process, but they have also worked to educate local officials who are unaware of the new policies. Advocates from LASVC recently succeeded in litigating over SB 946. In 2024, the city rescinded its "no-vending zones," which excluded vendors from high-traffic areas such as Hollywood. In a settlement, the city further agreed to cancel outstanding tickets and refund all fines paid by vendors who were charged with selling in these areas (Petrow-Cohen 2024). The settlement also requires changes to punitive law enforcement practices and sets up a process for vendors to have input in the design of new "special vending zones." Members of the LASVC continue to employ a combination of advocacy, litigation, and education to implement the hard-won policy changes described above.

Many cities and counties in California remain out of compliance with SB 946, and members of the LASVC are consulting with organizations in other cities as they push for more just street food vending policy in alignment with state law. In California, passing ordinances to amend local codes to comply with SB 946 is one easy step toward building a more just food system, but such changes still require active campaigns from local street food vending advocates. For example, in the small college town of Davis, California, vending was broadly illegal due to an ordinance that did not permit vendors to stop at any one location for more than five minutes. Local advocates began organizing to change the policy in 2016, but the City of Davis only removed their five-minute time limitation in 2023, well after the statewide legislation took effect. Local compliance is not automatic, and street food policy remains an afterthought for many decision-makers. With the state policy lacking oversight, street food legalization still requires advocates in each city to press for updates to municipal and public health codes.

Outside California, advocates and policy-makers can apply a similar framework to legalize street food vending by removing any restrictions on time, location, and equipment that are overly burdensome and more reflective of fixed food business interests and/or racist concerns about "neighborhood character," than of genuine public health concerns. Beyond simply lifting restrictions on street food vending, local and state governments can promote street food justice through smart policy. Following the lead of cities like New York, policymakers can establish programs to help street food vendors integrate with SNAP/EBT (Breck et al. 2015). They can offer incentives for vendors who offer healthy food options and for those who locate their operations in neighborhoods identified as having low food access. Finally, they can work with vendors to identify ways to streamline the permitting process or increase accessibility, such as by providing informational resources in multiple languages that reflect the diversity of the local vendor workforce.

Researchers also have a significant role to play in supporting street food justice. As noted above, relatively little scholarship engages with the topic of street food vending, and the literature overall is disproportionately focused on the issue of food safety. More research is needed to assess the wider public health impacts of street food vending, including the direct effects on food access in underserved communities and the indirect economic effects, both vendor wages and local business revenues. Researchers can build this knowledge by partnering with local groups to meet their identified needs. In Los Angeles, the "Sidewalk Stimulus" and "Unfinished Business" reports aided the LASVC's advocacy for policy change by building a counternarrative to push back on misconceptions, establishing consistent talking points about the scope of existing activities, and identifying specific barriers limiting street vending's contribution to the local food system. Researchers can also undertake impact assessments of vendor-based interventions in food deserts or local policy changes when these are set to occur.

Of course, public policy is only one arena for food system transformation. In Los Angeles, street vendors set their sights on policy change; having identified overpolicing and criminalization as the problem, they pursued city and state policies to decriminalize their work. As noted above, the policy outcome achieved in Los Angeles still relies on police; indeed, it included more funding for policing than it did for education about the new rules. With this contradiction in mind, food justice movement activists seeking to move beyond state-based solutions can build bridges and organize with street food vendors to incorporate them in a project of reimagining and recreating a (food) system that serves people rather than capital.

Street food vending involves not just nourishment and economic activity but cultural place-making. It is an expression of the right to the city for many otherwise marginalized residents, with benefits that ripple out to the wider public (Allison et al. 2021). Street food vending is also a highly adaptable extension of the food system that is viable for supplying food to poor communities and is an entrepreneurial opportunity for people needing alternative employment. As such, street food deserves serious attention in both the public health and food justice literatures, in urban and regional policy, and in food movement work, as a vehicle for the broader goal of building a healthy, inclusive, and sustainable food system.

References

Ababio, Patricia F., and Pauline Lovatt. (2015). "A review on food safety and food hygiene studies in Ghana." *Food Control* 47: 92–97.

Abello, Oscar P. (2022). "How street vendors finally made street food legal in California." *Next City*, October 4, 2022.

Agyeman, Julian, Caitlin Matthews, and Hannah Sobel. (2017). *Food Trucks, Cultural Identity, and Social Justice: From Loncheras to Lobsta Love*. Cambridge, MA: MIT Press.

Allison, Noah, Krishnendu Ray, and Jaclyn Rohel. (2021). "Mobilizing in the streets: The role of food vendors in urban life." *Food, Culture and Society* 24(1): 2–15.

Bennett, Cassidy, Joe Philipson, Scott Cummings, Doug Smith, Katie McKeon, and Brandon Payette. (2021). "Unfinished business: How food regulations starve sidewalk vendors of opportunity and what can be done to finish the legalization of street food." Public Counsel & UCLA School of Law Community Economic Development Clinic. https://law.ucla.edu/sites/default/files/PDFs/Academics/Unfinished_Business_Report_6_2021.pdf

Breck, Andrew, Kamila M. Kiszko, Courtney Abrams, and Brian Elbel. (2015). "Spending at mobile fruit and vegetable carts using SNAP benefits to pay, Bronx, New York, 2013 and 2014." *Preventing Chronic Disease* 12: E87.

Brinkley, Catherine, Benjamin Chrisinger, and Amy Hillier. (2013). "Tradition of healthy food access in low-income neighborhoods: Price and variety of curbside produce vending compared to conventional retailers." *Journal of Agriculture, Food Systems, and Community Development* 4(1): 155–169.

Brinkley, Catherine, Charlotte Glennie, Jose Flores and Benjamin Chrisinger. (2019). "'If you build it *with them*, they will come': What makes a supermarket intervention successful in a food desert?" *Journal of Public Affairs* 19(3): E1863.

Calderon, Karen B. (2015). "Street mobile food vending: Local policies, informality, and occupational health in immigrant communities." Masters of arts thesis, San Diego State University.

Donkor, Eric S., Boniface B. Kayang, Jonathan Quaye, and Moses L. Akyeh. (2009). "Application of the WHO keys of safer food to improve food handling practices of food vendors in a poor resource community in Ghana." *International Journal of Environmental Research and Public Health* 6(11): 2833–2842.

Elliott, Farley. (2015). *Los Angeles Street Food: A History from Tamaleros to Taco Trucks*. Charleston, SC: American Palate.

Faw, Brenda V., and Joyce L. Tuttle. (2014). "Mobile food trucks: California EHS-Net study on risk factors and inspection challenges." *Journal of Environmental Health* 76(8): 36–38.

Francis, Kaniyaa, and Catherine Brinkley. (2020). "Street food vending as a public health intervention." *Californian Journal of Health Promotion* 18(1): 1–16.

Hayden, Tiana B. (2021). "Street food as infrastructure: consumer mobility, vendor removability and food security in Mexico City." *Food, Culture & Society* 24(1): 98–111.

Lemon, Robert. (2019). *The Taco Truck: How Mexican Street Food is Transforming the American City*. Urbana: University of Illinois Press.

Li, Kathleen Y., Ellen K. Cromley, Ashley M. Fox, and Carol R. Horowitz. (2014). "Evaluation of the placement of mobile fruit and vegetable vendors to alleviate food deserts in New York City." *Preventing Chronic Disease* 11: E158.

Liu, Yvonne Yen, Patrick Burns, and Daniel Flaming. (2015). "Sidewalk stimulus: Economic and geographic impact of Los Angeles street food vendors." *Economic Roundtable*.

Los Angeles Street Vendor Campaign (LASVC). (2023). "About the Los Angeles Street Vendor Campaign." https://www.facebook.com/LAStreetVendorCampaign/about_details.

Martin, Nina. (2014). "Food fight! Immigrant street vendors, gourmet food trucks and the differential valuation of creative producers in Chicago." *International Journal of Urban and Regional Research* 38(5): 1867–1883.

Petrow-Cohen, Caroline. (2024). "In a win for street vendors, L.A. agrees to lift restrictions and cancel fines." *Los Angeles Times*, July 19, 2024.

Pringle, Peter. (2013). *A Place at the Table: The Crisis of 49 Million Hungry Americans and How to Solve It*. New York: Public Affairs.

Public Counsel. (2022). "Legislative alert: SB 972—Sidewalk food vending." Public Counsel. https://publiccounsel.org/wp-content/uploads/2022/11/Legislative-Alert-SB-972-1.pdf.

Yasmeen, Gisèle. (2001). "Stockbrokers turned sandwich vendors: The economic crisis and small-scale food retailing in Southeast Asia." *Geoforum* 32(1): 91–102.

Zepeda, Lydia, Anna Reznickova, and Luanne Lohr. (2014). "Overcoming challenges to effectiveness of mobile markets in US food deserts." *Appetite* 79: 58–67.

12 Digitizing Food Justice: How Community-Powered Tech Tools Are Being Used to Redesign Neighborhoods and Reimagine Food Justice

Antwi Akom, Tessa Cruz, and Analena Hope Hassberg

In March 2020, when schools, restaurants, and businesses of all kinds shut down and forced the world indoors, food access became a central concern. Both rural and urban US communities experienced shortages of food and household items (like sanitizing wipes and toilet paper) as fearful patrons panic-bought and hoarded supplies. In low-income urban communities and neighborhoods of color, where grocery stores are already scarce and overcrowded, lines stretched around city blocks, and shoppers waited hours, often leaving empty-handed. Food banks and mutual aid efforts were incredibly helpful but fell short of completely meeting community needs. Local and national groups that have historically organized around issues of food justice decried the ways that the COVID-19 pandemic exacerbated food insecurity and highlighted the structural inequalities that underpin access to healthy food. As we continue to adapt to the "new normal" wrought by the pandemic, new, powerful forms of activism are emerging that are shifting the food justice narrative, democratizing data and design, and digitizing the food justice landscape.

This chapter is about how food justice advocates are using digital maps and Black and Brown cartographies (a component of the powerful field of Black geographies) as powerful tools to fight against racism and food inequality and advance food justice. To illuminate this process and its potential, we focus on a digital mapping tool called Streetwyze. Streetwyze has been recognized nationally as a platform that successfully engages and empowers low-income and communities of color to affect positive neighborhood change (Akom et al. 2017; Akom et al. 2018). Developed originally for communities to share their experiences of the social and structural determinants of health locally, to date, the Streetwyze real-time community-driven data platform and process have been successfully implemented in at least thirty-five culturally and socioeconomically diverse communities in the US and another eight and growing worldwide (total participants approaching eight thousand, with over forty thousand data points collected) (Akom et al. 2024). The growing number of Streetwyze studies show that residents can effectively use mobile technology to record geo-tagged photos and share digital narratives and ratings that document the environmental and social aspects of their neighborhoods that either foster or hinder health and longevity, such as walking, parks, food access, housing, and other activities (Guan et al. 2023; D'Silva et al. 2021; Akom et al. 2016). These findings hold for residents of all backgrounds, including those from underresourced as well as more affluent communities (Holmes et al. 2021). Equipped with such data, residents can collaborate and work together to identify contextually relevant, low-cost solutions to promote their health and well-being, as well as the health and well-being of others in the local neighborhood.

In this way, the Streetwyze platform and process help build a culture of health and food justice by enabling vulnerable populations to collect relevant direct experience and information that is not otherwise accessible to public bureaucracies, local neighborhoods, or food systems. To improve health disparities and build a culture of health, communities must be viewed as assets and as key actors in transforming the social determinants of health. Streetwyze's real-time community-generated data increase the flow of critical information from communities to decision-makers while cultivating human capacities essential to putting solutions into action. Streetwyze helps communities more effectively use data to codesign food justice. Lessons learned from our previous work with grassroots organizations transforming food-insecure places into food-secure places show that when vulnerable populations are placed at the heart of efforts to build a culture of health, the data collection and story-mapping process itself becomes a health-promoting practice and a capacity-building activity.

Streetwyze improves communities'

1. capacity to put forward a vision of food justice and assert a set of community priorities that flows from that vision;
2. capacity to assess community vulnerabilities and assets and develop (or select) appropriate solutions based on a community's unique experience; and
3. capacity to build community voice and power to get priorities for food justice resourced and implemented.

When food justice planning processes are conducted without the community's capacity to envision and build power, they can become empty investments, simply producing a plan that sits on a shelf with little chance of being implemented. Thus, all three capacities are required to set the practice of building a community-driven culture of health into motion.

Digital Mapping as Counternarrative

Streetwyze and other digital mapping tools build on a long legacy of Black counter-mapping that emerged as forms of resistance against dominant discourses of structural racism (Alderman and Inwood 2021; Johnson 2022).

The Underground Railroad was led by Black cartographers such as Harriet Tubman and Ida B. Wells, who is well known for leveraging Black people's local knowledge ecosystems and lived experiences to fight against the pervasiveness of slavery and lynchings in American society (Alderman and Inwood 2021). Today, Black people aren't the only underrepresented group to utilize the power of counter-mapping and cartography. Indigenous communities, Latine communities, women, refugees, LGBQTIA+, and other POC communities have also utilized the power of maps to reimagine, redefine, and redraw their neighborhoods and experiences and gain access to institutional resources (Fraser 2017; Drakopulos and Lakhina 2018). However, Black Americans were among the earliest innovators of counter-mapping, deploying cartography as a form of resistance and engagement in the pursuit of designing liberatory spaces and places (Alderwood and Inwood 2021).

As slavery evolved into its afterlives, so did the forms of Black counter-mapping. During Jim Crow, Black Folks developed "green books" to help them navigate the landscape of racism and developed "wayfinding" aids to help negotiate racist realities and to cocreate visual maps that affirmed and celebrated Black life and culture happening beneath the veil. In his classic study, "The Philadelphia Negro," W. E. B. Du Bois produced maps that illustrated how Black folks had been excluded from institutional resources and visualized how many were living in abject poverty (Du Bois 1899). The Black Panther Party continued the practice of Black counter-mapping in their efforts to illuminate resistance, agency, and political contestation (Hassberg 2020). They used maps and gathered over fifteen thousand petitions to create new police districts in Berkeley and San Francisco, largely along racial lines, to protect the people in the community (Tyner 2006).

Building from the history of Black and Brown cartographies, we argue that today's food activists are using digital mapping technologies to advance food justice. This line of research contrasts with what have become the dominant, dystopic perspectives on race and technology. Anna Everett's groundbreaking work, "The Revolution Will be Digitized: Afrocentricity and the Digital Public Sphere," critiques "the overwhelming characterizations of the brave new world of cyberspace as primarily a racialized sphere of Whiteness; that consigns Black people to the low-tech sphere—when they are present at all." (Everett 2002). Pushing even further, Ruha Benjamin describes a "New Jim Code" in which tech designers encode racial, gendered, and socially biased judgments into next-gen AI systems, but claim the racial and social bias results are entirely neutral (Benjamin 2019).

While agreeing that the harms of technology and artificial general intelligence (AGI) will disproportionately impact poor people, women, and POC; our chapter offers an alternative perspective on the ways that BIPOC digital haves and digital have nots are designing, building, and leveraging community-driven, community-powered, and community-owned tech tools and platforms for social good. We intentionally highlight the flip side of the "digital caste system" argument by exploring how community-driven, community-powered, and community-owned digital platforms can be used as tools of resistance, agency, and political contestation. Benjamin's Ida B. Wells Just Data Lab at Princeton works in this same vein, though not on food.

Indeed, there is a dearth of information on how food justice activists are leveraging technology tools and frameworks that better capture how food matters are spatial matters, and spatial matters are racial matters (Garth and Reese 2020). This technology gap in the food justice literature needs to be filled to better articulate, capture, and comprehend how participatory technology and civic engagement tools and platforms are helping food activists democratize data and fight for racial equity to advance food justice. In particular, this chapter is about how Streetwyze is helping food activists "flip the script" and move from a focus on Big Data to those voices at the edges: those people, places, and spaces who are most deeply impacted by structural racism and systemic injustice. Our chapter investigates how BIPOC communities are transforming food inequality into food sovereignty, by exploring how Black and Brown food justice activists are utilizing technology to play a positive role in impacting community development and are digitally designing new innovative ways to achieve food justice.

Our research shows that if we listen closely to the voices at the margins and analyze how they are leveraging information and communication technologies in terms of resistance, agency, and political contestation—then we can begin to make visible the invisible ways that Black, Indigenous, and POC (BIPOC) communities are leveraging tech tools to create food access and food agency and overcome food apartheid to cocreate food justice and food liberation (Akom et al. 2016).

As the vignettes that follow will demonstrate, Streetwyze and other digital mapping tools build on this legacy by enabling communities to cocreate real-time representations of their lived experiences. In the past several decades, researchers in public health and the social sciences have sought to share power over the research process with the communities they study through youth participatory action research and community-based participatory research. But even research guided by these methodologies can often remain university-led, as was initially the case with the project described in chapter 14. By combining the democratizing instincts that animate participatory approaches with real-time, community-powered and community-controlled technologies, digital countermapping efforts can help us dream toward an entirely new conceptual and methodological approach that we call "reality methodology."

Reality methodology builds on Edmin's (2016) concept of reality pedagogy, guiding teachers to bring students' everyday realities into the classroom. Reality methodology enables communities to use digital tools like Streetwyze to represent their own lived experiences in a way that can challenge power and create more liberated worlds. In this sense, reality methodology becomes a research and learning approach that focuses on understanding places and spaces based on the reality of the experiences of people living there in real time. Beyond participatory action research and community-based participatory research approaches, it has the potential to ensure equitable interpretation, representation, and accountability to communities that have historically been locked out, left out, undervalued, and underappreciated in data sharing in research and design processes (Costanza-Chock 2020; Safir and Dugan 2020).

The Streetwyze App: Digitizing Food Justice

Our approach synthesizes work from science and technology studies, critical race theory, critical geographies, critical mapping, designing justice, and decolonization literature with the experiences of everyday people, specifically grounding their food justice work from the perspective that structural racism produces stress, income inequality, lack of access to institutional resources, and food inequality. What is novel about our work is the ways we shed sunlight on how BIPOC communities are leveraging technology to build power and self-determination so that we can all collectively experience a more just and equitable food landscape and begin to end food apartheid through the power of community-driven design and community informatics.

To combat structural racism and food inequality, we introduced the Streetwyze model in the food justice space in 2016 when we collaborated with a group of students in East Oakland to transform a so-called food desert into a food oasis (Akom et al. 2016). Many of the students in our class were from Oakland's poorest neighborhoods—places and spaces where people die early from unnatural causes and suffer from illnesses that better diets may delay or prevent altogether, including high blood pressure, diabetes, and cardiovascular disease.

The students we worked with wanted to test whether their neighborhood was a food desert or a food oasis. British and Scottish researchers introduced the concept of food deserts in the mid-1990s, suggesting a link might exist between the distance to a grocery store and the diets of poor people. To test our hypothesis inside and outside the classroom walls, we gathered around a few small computers and looked at all the publicly available data that indicated that the neighborhood the students lived in was a food oasis—a place with more grocery stores than liquor stores and corner stores. This did not mesh with the students' lived experiences (Cummins 2002). They knew from living in the neighborhood that those so-called grocery stores displayed on public maps as official knowledge were not accurate at the street level in terms of street data, and they were determined to do something about it (Safir and Dugan 2020).

What was novel about our work in 2016 was that our project aimed to unpack the story of place using the power of technology, real-time data, community-driven data, and data visualization in an effort to better understand the place-based mechanisms that produce food equity and food disparities with our nation's most vulnerable populations so they no longer feel "stuck in place" (Sharkey 2013).

During the Streetwyze onboarding process, we also introduced students to Africana studies, ethnic studies, women's studies, sexuality studies, critical geography, critical race studies, and data science (in relationship to science and technology) and trained them in reality pedagogy and methodology.

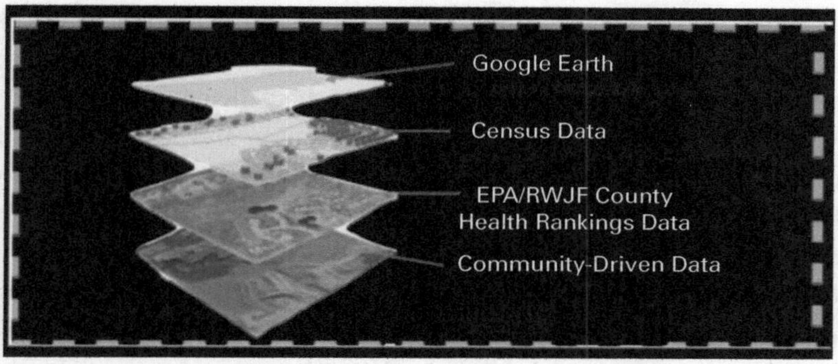

Figure 12.1
Example of Streetwyze data layers—integrating big data with community-driven data. Image courtesy of Streetwyze.

We asked the students, (1) What kinds of food outlets are available within this neighborhood? (2) What is the nature of East Oakland's food environment and how does it vary by the racial composition and population of neighborhoods? and (3) How can community-driven and community-powered asset-based community development initiatives help reimagine the food landscape here? We centered the concept of asset-based community design with the students because it demonstrates how community members have power as active, autonomous agents in their own lived experiences and, in spite of structural racism and other forms of oppression, struggle to improve the social and material conditions in their own lives (Akom 2011).

To ground-truth what was happening in the neighborhood, we introduced students to the Streetwyze platform and process.[1] The Streetwyze mobile, mapping, and SMS platform is a community-driven, community-powered, community-owned research model founded by the nonprofit Dope Labs—which stands for Digital Organizing Power-Building and Engagement. Community members use Streetwyze to filter data into three categories: "good stuff," "bad stuff," and "fix stuff," which allows community participants to share and map amenities and real-time experiences. They can call attention to places that might not be visible and places that bring them joy or perhaps are causing harm or negativity in the neighborhood, and they can share ways that community institutions and establishments can improve to meet the needs of local residents better so we can have upstream solutions to social determinants of health (figure 12.1).

Streetwyze is a scalable, sustainable, "bottom-up" process and platform that engages, empowers, and enables Black, Indigenous, and POC activists, street scientists, and students to (1) gather and share their own real-time experiential data to power local change; (2) build power and self-determination and increase community voice through digital storytelling; (3) make invisible assets visible and engage in asset-based community development processes and place-based development; and (4) democratize data through our mobile platform, where anyone with a cell phone can become a videographer and move the crowd (Corburn 2005). Overall, the results of our work in East Oakland were powerful. From the data collected and presented to local city council meetings and school administrators, the local high school was able to start a weekly farmers market on campus, which increased access to fresh food. Additionally, the CalFresh Program helped facilitate the transformation of some of the liquor stores in the neighborhood by offering funds to owners for more refrigeration units so they could carry more fresh, culturally appropriate, and affordable food and store fresh food longer. Finally, the data were used to help gather support for Measure JJ, a multimillion-dollar bond measure that created a local food commissary and community garden to serve the Oakland Unified School District.

Counter-Mapping During COVID

Little did we know in 2016 that the valuable lessons we learned about digitizing food justice would have a major impact in 2020 when COVID began spreading across the US and physical, public, and social infrastructures began to shut down. Digital infrastructures—those infrastructures that help people struggle, survive, and find access to life-saving resources across spaces and places—began to take on much more significance with respect to food justice and civic and social life (Zuckerman 2020). During COVID, we heard from many people that they weren't sure where their next meal would come from and that the weight of that uncertainty—the waiting, the not knowing, the stress, the grind, the grit, the fatigue—was literally eating at them physically, mentally, and socially. We heard from young people that they were often embarrassed to share their real hunger needs with their friends and that many were having challenges focusing in the classroom. Given the sheer life and death situation for many, the Alameda County Community Food Bank (ACCFB) invited Streetwyze to support their program seeking to feed those underresourced neighborhoods hit hardest by the pandemic with a pilot food justice intervention program aimed at connecting food-insecure people to life-saving services to increase equitable access to nutritious, affordable, and healthy food.

Our partnership with ACCFB aimed to implement a community-driven mapping campaign that would achieve the following goals: (1) better understand the community experience of different food banks across the ACCFB network, gathering real-time data on invisible qualitative measures that are important for all income levels, such as being treated with dignity and respect, documenting and recording wait times (because everyone's time is valuable), quantity and healthfulness of food, satisfaction, and trust; (2) identify the alternative food outlets and sources outside the ACCFB network that communities are leaning on during COVID, mapping them geospatially and uploading digital narratives of food access, food waste, food conservation, and sharing; and (3) build a digital peer-to-peer green book food wayfinding system to increase access to fresh, affordable, locally grown produce for low-income communities and communities of color by utilizing data-driven strategies to improve programming, distribution, and access.

To achieve these goals, the Alameda County Community Food Bank shared efforts from around the country trying to better understand structural racism and equitable food access. These included Feeding America's Identifying Racism Dashboard (Feeding America 2021) and Chicago's Food Equity Agenda (Advancing Food Equity in Chicago 2023). The digitizing of food-based movements is important because, according to the Pew Research Center (2021), 97 percent of Americans own a mobile phone, and for those that don't, the federal government provides eligible low-income individuals with free phones and unlimited texting. This means that digital access continues to be on the rise in the US, and the next generation of food justice must include accessible technology to fuel robust community-driven movements.

During our project period, ACCFB and the Streetwyze team trained over 75 community participants, who mapped over 250 food resources on the Streetwyze mapping platform. ACCFB then utilized these data to improve current programs, create new initiatives to better address the ground-level food needs of BIPOC community members within Alameda County, and explore new partnerships with other food distribution centers that BIPOC communities were leaning on during the pandemic, including schools, community organizations, churches, clinics, and more.

Mapping the Barrio: Digitizing Food Justice in San Diego

In 2019 a small group of Latine women gathered in kitchens to discuss how to create healthier food traditions to change communities in San Diego. Kitchens as spaces of resistance are an overlooked aspect of the food justice movement (Kohl and McCutcheon, 2015). However, many "kitchen table talks" have sparked the seeds of neighborhood resistance and neighborhood improvement. This is especially true in San Diego, where many Latine neighborhoods have deep roots tracing back to the early twentieth century when Mexican laborers settled in the area working in the nearby shipyards and canneries and utilized traditional foods and culture to thrive in the face of systematic oppression and structural racism. As a result, there is a strong sense of Latine history, cultural pride, and activism baked deeply into many San Diego neighborhoods like Barrio Logan, Chicana Park, Logan Heights, City Heights, and San Ysidro, to name a few. It is visible and palpable through numerous art galleries, mural-lined streets, and social justice organizing. Along with this vibrant history, US Census data show that the median household income for Hispanics and Latinos living in San Diego County is 27.5 percent lower than that for the rest of the county, and the unemployment rate is 1.5 percent higher than the national average.[2] These inequalities are also reflected in the differential access to good-quality health and food infrastructures.

To address and overcome these structurally racist health inequalities, a small group of women from Olivewood Gardens who called themselves the Kitchenistas partnered with a coalition of ten grassroots multicultural community-based organizations, including Promotoras, Dope Labs, Streetwyze, and the University of California–San Diego, to develop programmatic and policy solutions for the food apartheid impacting the San Diego community. They named the project HEALthy 4 You (H4Y) and used a "train-the-trainer" model to map ground-level stories about adverse childhood experiences, childhood obesity, and the social determinants that affect individual and community health.[3]

The Kitchenistas and other partners invited Streetwyze into H4Y because Streetwyze is a multilingual, multimodal platform, meaning that in addition to using Streetwyze in native languages (the platform is currently available in Spanish, Vietnamese, Korean, Chinese, and Arabic), communities are able to upload information in a variety of different formats (text, images, video, and audio), which is important for lower-literacy and monolingual non–English speaking communities. Having an easy-to-use platform was key for them sharing their unique voices and experiences living in neighborhoods inundated by fast-food restaurants and convenience stores with few options for fresh produce and other healthy foods.

The ongoing Kitchenista-Streetwyze project has been very successful. So far over 1,500 community participants have mapped over 7,500 data points across San Diego County, and overwhelmingly, most data points mapped were about food. For instance, when asked about "bad stuff," places that contribute negatively to their health, community members in Julian, California, called out the lack of affordable and accessible grocery stores and the impact of limited public transportation on their ability to access food. Similarly, community members in National City, California, shared disappointment with the quality, selection, and price of food available to them, as well as unsanitary conditions and food safety violations in restaurants.

When asked about "good stuff," places that contribute positively to their health, Julian residents commended local marketplaces with fresh and nutritious groceries, family-owned food businesses with attentive staff and good customer service, and local restaurants offering convenient delivery services and extended hours. They also mapped food drives and local pantries, as well as schools that provided free food for students on weekends. In National City, the community praised locally owned shops that sell pesticide-free fruits and vegetables and long-standing cultural institutions that offer culturally authentic food, which many noted are important pillars of hope in the face of rising housing insecurity, homelessness, and gentrification.

The "fix stuff" data points across communities emphasized a desire to see new farmer's markets and grocery stores where residents would have consistent access to produce, meat, dairy products, and vegetarian/vegan options. There were also calls for better access to food resources such as CalFresh and a wider selection of certified organic foods to accommodate dietary restrictions and preferences.

Stories like the ones mapped in San Diego County helped increase community pride around the places that upheld the cultural food fabric of the neighborhood, increased awareness of new food resources through peer-to-peer learning, and helped amend or refute Big Data sets (such as Google Maps) that may misrepresent neighborhood foodscapes, for instance, by listing incorrect operating hours or failing to note whether a store takes WIC or EBT benefits. Furthermore, through the train-the-trainer model, Kitchenistas, community members, Promotoras, and community health advocates increased their tech literacy and developed new skills to support their movement and capacity-building efforts with community residents on the ground so grassroots movements can be better sustained, maintained, scaled, and documented over time with the support of digital tools and resources like Streetwyze. As a result of systemic inequities, National City has one of the highest rates of obesity and diabetes in San Diego County, but Streetwyze helped the Kitchenistas and other community organizers band together to demand access to healthy food and to clean and safe parks, improve exercise and recreation opportunities for Latine youth, and advance access to health care and housing.

Conclusion

As authors, we've all been labeled as "street kids" at one time or another in our lives, and far too often, this term has been weaponized in derogatory ways toward BIPOC communities. All of us, to varying degrees and at different phases of our lives, have been looked down upon for being from the hood, ghetto, Barrio, Rez, Block, corner, or some derivative of something negative, unwelcomed, and unwanted.

This perception replicates itself in food production and distribution systems. People from certain communities, neighborhoods, and towns and from the street are perceived as bad and thus deserving of fast food, liquor stores, and other unhealthy substances. Our historical cultural foodways are also seen as deficient and are blamed for poor health outcomes.

Our chapter begins to counternarrate the concept of street wisdom as a place of power and possibility by demonstrating how the Streetwyze platform and process not only help to digitize the food justice movement but also help to build a new approach that we term reality methodology. This approach weaves together reality pedagogy's focus on everyday lived realities with participatory technologies, using digital tools to democratize design and data creation. Reality methodology stands out for its innovative integration of mobile technology, emerging technologies (including AI), community informatics, geospatial tools, and social media, in ways that enable the collection of real-time lived experiential data that are community-driven, community-powered, and community-controlled. An additional distinction between reality methodology and community-based participatory research is that it is not top-down and university or corporate controlled; instead, reality methodology is bottom-up and community-led from problem identification, data collection, and analysis, through to dissemination. With reality methodology, everyday people in local neighborhoods and communities are the key experts, providing the lived expertise needed to both ground truth and enable communities' street-level realities to guide social service provision, policy, and mutual aid to support future community-led research, organizing, advocacy, and action. We have shown how this new approach can highlight the role that participatory technologies and methodologies can play in positively impacting asset-based community development and protective resiliency factors that impact food justice and community health and well-being.

We have shown how this methodology enables communities' street-level realities to guide social service provision, policy, and mutual aid. Reality methodologies and real-time technologies that are built for us and by us (FUBU) begin to challenge the New Jim Code (Benjamin 2019). While agreeing that a digital caste system exists, our chapter has explored how Black and Brown communities are transforming themselves from consumers to producers and building underground tech tools that can be used for resistance, agency, food justice, and liberation.

In this chapter, we have lifted up and shared how community groups can build, develop, and leverage their own tech tools to power food justice movements toward more inclusive outcomes. However, we do not believe that Streetwyze (or any other technology for that matter) is a panacea. As Safiya Noble (2018) brilliantly articulates, "An app alone will not save us." Instead, to build a movement to counter the New Jim Code, we introduced the concept of digitizing food justice to help us better understand how cocreating community-driven, community-powered, community-owned technologies and reality methodologies can enable BIPOC, rural, and disability-forward communities to transform the social and material conditions that oppress us so we can move from surviving to thriving and have access to institutional resources that help people get free. This is a first step, and we invite all of you who feel the call to build upon this work. We are hopeful it will grow.

Notes

1. Ground-truthing pairs community narratives with quantitative data sets to more accurately define the ways that real people are actually experiencing their neighborhoods on the ground.

2. U.S. Census Bureau, "Median Household Income: 2015–2019"; U.S. Census Bureau, "American Community Survey 1-Year estimates. 2019: Employment Status San Diego County and U.S.," https://data.census.gov/.

3. Adverse childhood experiences (ACEs) are traumatic events such as abuse, neglect, parental substance use, incarceration, and domestic violence that occur before a child reaches the age of eighteen. Experiencing ACEs can impact lifelong health outcomes.

References

Akom, A. A. (2011). "Black emancipatory action research: Integrating a theory of structural racialisation into ethnographic and participatory action research methods." *Ethnography and Education*, *6*(1), 113–131. https://doi.org/10.1080/17457823.2011.553083.

Akom, A., Cruz, T., Ailkon, A., and Agyeman, J. (2025). *Digitizing Food Justice: Nurturing Food*. Routledge Press. Forthcoming.

Akom, A., Shah, A., and Cruz, T. (2017). "Streetwyze: mapping 2 mobilize & people powered place-making toolkit. Streetwyze. https://www.streetwyze.com/wp-content/uploads/2021/01/StreetwyzeToolkit.pdf.

Akom, A., Shah, A., and Cruz, T. (2018). "Building smart, resilient communities and schools: Using a technology platform to transform the social determinants of education with our nation's most vulnerable populations." *Green Schools Catalyst Quarterly*, *5*(3), 22–39.

Akom, A. A., Shah, A., and Nakai, A. (2016). "Kids, kale, and concrete: Using participatory technology to transform an urban American food desert." In P. Noguera, J. Pierce, and R. Ahram (eds.), *Race, Equity, and Education*. Springer. https://doi.org/10.1007/978-3-319-23772-5_4.

Akom, A., Shah, A., Nakai, A., and Cruz, T. (2016). "Youth participatory action research (YPAR) 2.0: How technological innovation and digital organizing sparked a food revolution in East Oakland." *International Journal of Qualitative Studies in Education*, *29*(10), 1287–1307. https://doi.org/10.1080/09518398.2016.1201609.

Alderman, D., and Inwood, J. (2021). "Memory-work in Montgomery, Alabama." Focus on Geography. https://www.focusongeography.org/publications/photoessays/alderman/index.html.

Benjamin, R. (2019). *Race after Technology: Abolitionist Tools for the New Jim Code*. Polity.

City of Chicago. (2023). Advancing Food Equity in Chicago. https://www.chicago.gov/city/en/sites/advancing-food-equity-in-chicago/home.html.

Corburn, J. (2005). *Street Science: Community Knowledge and Environmental Health Justice*. MIT Press.

Costanza-Chock, S. (2020). *Design Justice*. MIT Press.

Cummins, S. (2002). "'Food Deserts': Evidence and assumption in health policy making." *BMJ, 325*(7361), 436–438. https://doi.org/10.1136/bmj.325.7361.436.

Drakopulos, L., and Lakhina, S. J. (2018). "How 'story maps' redraw the world using people's real-life experiences." The Conversation August 10. https://theconversation.com/how-story-maps-redraw-the-world-using-peoples-real-life-experiences-98051?utm_source=clipboard&utm_medium=bylinecopy_url_button.

D'Silva, J., J. Moze, J. H. Kingsbury, R. K. Lien, C. M. Matter, B. Brock, and A. Akom. (2021). "Local sales restrictions significantly reduce the availability of menthol tobacco: Findings from four Minnesota cities." *Tobacco Control, 30*(5), 492–497.

Du Bois, W. E. B.. & Eaton, I. (1899). *The Philadelphia Negro: A Social Study*. Philadelphia: Published for the University.

Everett, J. (2002) "Organizational research and the praxeology of Pierre Bourdieu." *Organizational Research Methods* 5 (1): 56–80.

Feeding America. (2021). "Interactive Data." https://www.feedingamerica.org/research/interactive-data.

Fraser, J. A. (2017). "Why Amazonian forest peoples are "counter-mapping" their ancestral lands." The Conversation. September 26. https://theconversation.com/why-amazonian-forest-peoples-are-counter-mapping-their-ancestral-lands-84474?utm_source=clipboard&utm_medium=bylinecopy_url_button.

Garth, H., and Reese, A. M. (eds.). (2020). *Black Food Matters: Racial Justice in the Wake of Food Justice*. University of Minnesota Press. https://doi.org/10.5749/j.ctv182jtk0.

Guan, A., Cruz, T., Sowell, J., Mattias, B., Shah, A., Hope Hassberg, A., Shariff-Marco, S., Akom, A., and DeRouen, M. C. (2023). "Combining mixed methods and community-based participatory research approaches to identify neighborhood-level needs during the COVID-19 pandemic." *Journal of Mixed Methods Research*. December 15. https://doi.org/10.1177/15586898241300785.

Hassberg, A. "Nurturing the revolution." (2020). In H. Garth and A. M. Reese (Eds.), *Black Food Matters: Racial Justice in the Wake of Food Justice*. University of Minnesota Press.

Holmes, L. M., McQuoid, J., Shah, A., Cruz, T., Akom, A., and Ling, P. M. (2021). "Piloting a spatial mixed method for understanding neighborhood tobacco use disparities." *Social Science & Medicine, 291*, 114460. https://doi.org/10.1016/j.socscimed.2021.114460.

Johnson, P. (2022). *Unspoken Maps: How Black Women Principals Experience and Respond to Anti-Black Linguistic Racism*. Dissertation. San Francisco State University.

Kohl, E., and McCutcheon, P. (2015). "Kitchen table reflexivity." *Gender, Place and Culture, 22*(6), 747–763.

Noble, S. (2018). *Algorithms of Oppression: How Search Engines Reinforce Racism.* NYU Press.

Pew Research Center. (2021). "Mobile fact sheet." Pew Research Center. April 7. https://www.pewresearch.org/internet/fact-sheet/mobile/.

Safir, S., and Dugan, J. (2021). *Street Data: A Next-Generation Model for Equity, Pedagogy, and School Transformation.* Corwin.

Sharkey, P. (2013). *Stuck in Place: Urban Neighborhoods and the End of Progress Toward Racial Equality.* University of Chicago Press.

Tyner, J. A. (2006). ""Defend the ghetto": Space and the urban politics of the black panther party." *Annals of the Association of American Geographers, 96*(1), 105–118. https://doi.org/10.1111/j.1467-8306.2006.00501.x.

Zuckerman, E. (2020). "What is digital public infrastructure?" Center for Journalism & Liberty. https://www.journalismliberty.org/publications/what-is-digital-public-infra structure.

13 Decolonized, Community-Rooted™ Food Justice: The Process of Shifting Power and Decision-Making in Food System Planning Processes

Kamal Bell, Natalie Eley, Vivette Jeffries-Logan, Justin Robinson, Camryn Smith, Gizem Templeton, Gretchen Thompson, and Jen Zuckerman

Imagine Any Community

Two white men are chatting in the hallway of an esteemed academic institution—well-regarded leaders in their fields of expertise, one a lifelong academician and one with political aspirations, both with a desire to leverage their knowledge to do good work.

An idea arises: "Let's make our city a model food community! We could get all the policies right here, then share that model out broadly." Through relationship and a long track record with philanthropic organizations, the men get funding. The academician hires a white woman as the project lead. The politician calls for a public announcement of the idea. They launch the project.

In the community, no one knows this is happening. No one asked for it to happen. Despite this reality, the project begins. And so did our story.

Our story is also where this narrative shifted, with a process of harm and healing that evolved into a different way of being. The story is not a formula or a model, but a narrative of our institution-initiated "food system solution" and how it evolved into a decolonized process that laid the foundation for systemic change in our community. Similar to Bellante et al. (see chapter 15), we offer our story as a mechanism for preserving historical activism, challenging dominant narratives in the food industry, and amplifying the work within BIPOC communities to envision and realize a more equitable and just future for food.

Here, we detail our emergent process (figure 13.1), highlighting individual experiences from initial outreach to the dissolution of the formal effort, cut

---

This chapter has equal authorship and is ordered alphabetically.

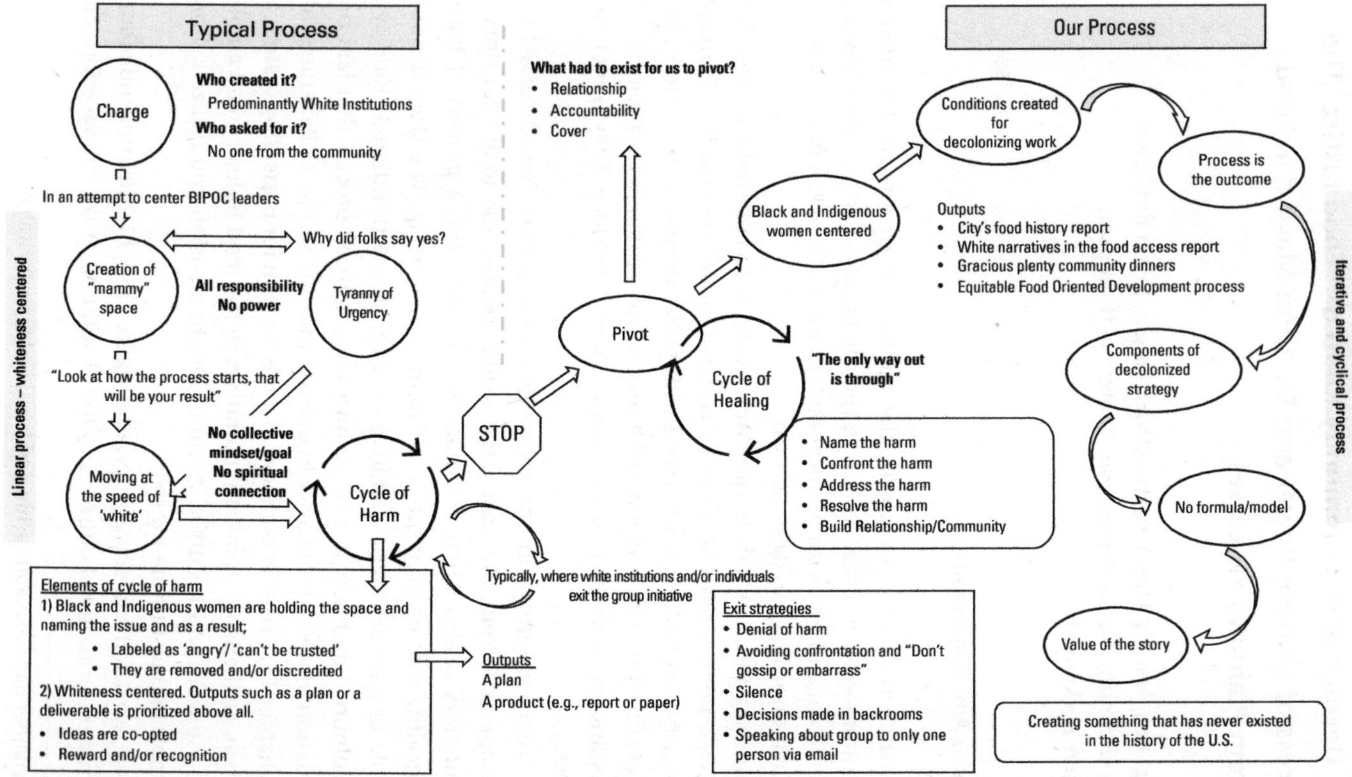

Figure 13.1

Diagram describing the process that emerged in this collective compared to a more typical process. Image courtesy of the authors.

short by COVID-19 lockdowns. Individual narratives detail personal histories and the social positionality of our team to offer a racialized, gendered, and power perspective on the process as well as highlight incidents that caused harm or promoted healing, for the purposes of elucidating the dynamics of convening multiracial teams to address food justice. Our analysis addresses four key themes:

1. Centering Black and Indigenous leaders and the role of predominantly white institutions (PWIs)
2. Differentiated impact of the burden of leadership for Black and Indigenous women
3. Emergent processes of harm and healing
4. Navigating systems of power and privilege within PWIs

We are specific with language in this narrative. Our working definitions for key terms necessary to understand, write about, and analyze food justice work are defined as follows:

Predominantly white institutions are traditionally defined as a college or university where 50 percent or more of the students are white (Bourke 2016). Here, we extend this label to other institutions such as nonprofits, local government, and businesses that are white-founded and center whiteness.

Whiteness refers to the power of racialized whiteness in "the way that white people, their customs, culture, and beliefs operate as the standard by which all other groups are compared. The normalization of white racial identity throughout America's history has created a culture where nonwhite persons are seen as inferior or abnormal. White dominant culture operates as a social mechanism that grants advantages to white people, since they can navigate society both by feeling normal and being viewed as normal" (National Museum of African American History and Culture n.d.).

White-centered speaks to whiteness as "a dominant cultural space with enormous political significance, with the purpose to keep others on the margin. . . . White people are not required to explain to others how 'white' culture works, because 'white' culture is the dominant culture that sets the norms. Everybody else is then compared to that norm. . . . In times of perceived threat, the normative group may well attempt to reassert its normativity by asserting elements of its cultural practice more explicitly and exclusively" (Estable et al. 1997, 21).

Community Rooted™ is a lived experience framework developed by Communities in Partnership to differentiate between paternalistic, charity-based nonprofits and organizations and organizations like Communities in Partnership who approach "community development to amplify the lived experience of residents and to address challenges around accountability and decision-making experienced by some community-based organizations. The approach utilizes governance structures, decision-making practices, and strategies that reinforce accountability, enhance community ownership, and emphasize systems-level change. A community-rooted organization may be nonprofit, for-profit or faith-based, formal or informal. The key is that its mission, vision, and day-to-day operations center on those most affected by community conditions" (Smith et al. 2020).

Decolonized/decolonizing is "the active resistance against colonial powers, and a shifting of power towards political, economic, educational, cultural, psychic independence and power that originate from a colonized nation's own indigenous culture. This process occurs politically and also applies to personal and societal psychic, cultural, political, agricultural, and educational deconstruction of colonial oppression" (Racial Equity Tools, Glossary n.d.).

In putting our story on paper, we talked about our experiences as a group, reflecting on events that punctuated our work, the overall process, and our collective experience. We wove our reflections into the narrative presented in this chapter, with highlights from individuals attributed by name. Our group of eight coauthors includes individuals who identify as Black, white, Indigenous (Occaneechi Band of the Saponi Nation), and immigrants and who work in nonprofit, university, and community settings. Each of us with our complex identities entered and worked in our collective space and contributed to this presentation of our experience.

"How You Start Is Where You Typically End."

We mean well, those of us working at PWIs. We are driven by good intent and fall back on our good intent at each stage of our process. What good intent translates to is that from the outset, whiteness is centered and protected. That pattern launched our story.

The good intent was to make our city a model food community. As the project lead, Jen, a white woman who has spent her career working in PWIs, shared, "It sounded 'right.' Get all the policies and practices aligned for good food in one community? I believed that this would benefit everyone, so I wanted to make it happen."

We never considered to ask "Whose model? And, for whom?" Because in PWIs we move from a white-centered, universalist framework; we assumed that "best practice" meant best practice for everyone, so it was simply a matter of bringing all those best practices to bear. The process took shape as expected—a linear process of gathering information, presenting it to stakeholders, choosing priorities, and creating a final product—in this case, a plan.

This process included the desire to center community-rooted™[1] leaders as the stakeholders. Community-rooted™ leaders live in and are in relationship with the communities they work with and are differentiated from community-based organizations, which typically are founded and operated by individuals outside the communities they serve. Gretchen, the white woman evaluator, recalled, "I remember saying that we needed to engage BIPOC [Black, Indigenous, and people of color] communities through our work before going any further. I was excited about adding a process evaluation to the work so that experiences could be documented, and 'best-practices' could also be shared."

Gretchen invited Jen to learn about a community-centering methodology developed and used by her organization. The training framed the concept of a core "design team," a small group of community-rooted™ leaders who drove the content of the project, reporting back to a "collaborative council" made up of representatives from PWI nonprofit and government organizations. The intent driving this action was to center Black and Indigenous leadership, but the action was misaligned. It provided the optics of Black and Indigenous leadership but, in fact, continued to center whiteness, as the goal of the project had been set by the PWI, and the design team, while driving content was accountable to a council of PWIs.

Missing from the design was a reparative strategy. As a Black community-rooted™ woman, Camryn noted, "If you are well intentioned, immediately start giving over reparative means to people from the onset. You have to trust people that are directly impacted, and you have to cover them." Providing cover means that well-intentioned people should, in addition to providing financial resources, shift power and move accountability to the BIPOC people they have asked to lead. This accountability entails leveraging the power and privilege of whiteness and/or institutions to deflect and shield communities of color and activists seeking to create equitable change from people within white supremacist systems that would seek to undermine any power-shifting strategies.

Absent a reparative strategy, our project continued its linear path: creating the space that PWIs replicate time and again—giving Black and Indigenous people all the responsibility and none of the power. As a Black man invited into this space, Justin referenced the construct stemming from Black feminist theory of a "mammy space," a construct specific to Black women, which depicts them as faithful, obedient servants bearing all the responsibility of domestic service and being subordinate to elite white male power. The mammy space construct is analogous here in that we created a similar power dynamic (Collins 2022). Kamal, a Black male farmer related his experience: "You just put people together and then expected a result. That's not fair to us, because our experience is so different. It's hard to do decolonizing work when literally that's how colonization works—you just get thrown into a situation and you have to adapt."

The pattern created in this project is the same pattern that is repeated across communities and projects, time and time again. So, why do community-rooted™ leaders often feel obligated to join? It is the tyranny of urgency, the "constant tension between the urgent and the important" (Hummel 2009). Camryn recalled, "It's the power differential. I had that pressure on me because I knew this was a way for us to have a best practice that was directly connected to other Black-led and Indigenous-led organizations . . . and I needed to make that move."

The project charged ahead with the assumption from Jen as the project lead that power and equity issues had been addressed because Black and Indigenous leaders were present in the space. Qualities endemic to the culture of PWIs took over, setting up an activity series that sought only to achieve the project outcome, indifferent to the impact on the people involved. The "stakes" from the perspective of the PWI were technical, not relational. Those stakes included meeting the stated project objective of "making the city a model food community" by producing a plan, in writing, within the timeline and the budget established by the funding entity. Camryn called this stage "moving at the speed of white." She explained, "White institutions want to move at the speed of light, not realizing where they are in terms of their own competency. This stuff takes time. When you're talking about equity, you need to slow down. What needs to happen is a very basic power transfer."

The perfect storm was brewing. A group existed, moving out of obligation to a goal established by an external entity and without shared understanding of the root causes of the problems. As Kamal noted, there was "no collective understanding" in the group of how we want to interact and decidedly "no spiritual connection" for or collective analysis of the space we were in together. Jen recalled, "I made assumptions that because folks said yes and showed up, that we were on the same page. I brought in a contractor, a white male, who had a track record of producing the kind of plan I believed we were moving towards, because that is what I understood I needed to achieve as a result of this work. What I didn't understand was the wildly divergent understanding, responsibilities, and motivations among folks in the room."

In addition to a difference of analysis and understanding in starting this effort, this divergent experience was also reflected in the compensation, contracts, and contract process for the consultant and the design team participants. The white consultant was brought in as the project manager and was paid a fee for his business services. His contract fee was not questioned or scrutinized by the PWI. Conversely, the design team community participants were paid an individual consulting rate that was lower than that for the white consultant's services. Even with this lower rate, because the fee was a consultant rate and not a community stipend rate, their contracts were scrutinized by the PWI—a process that we note here further contributes to the extraction and appropriation of intellectual property from Black and Brown communities, something we discuss further below. The act of advocating for a consultant fee for community leaders as opposed to a stipend was a breach of PWI protocol, creating tension within the PWI that Jen as the project lead would then have to navigate throughout the remainder of the project.

The momentum of this process—led by PWI and a white male contractor hired by that PWI—moved into a cycle of harm. The team participated in an extensive racial equity training series, at the advice of Vivette, a community racial equity leader/facilitator. Vivette and the design team, which included the white male facilitator and white representatives of PWIs (Jen and Gretchen), developed a set of guiding principles out of the training experience. The guiding principles codified that a decolonized approach is foundational to our work, including the willingness to course-correct, have flexible timelines, give attention to relationship building, address conflict and discomfort, and elevate the voices and power of people of color–led organizations, collaborators, and communities.

Jen held the false assumption that experiencing training as a team and establishing shared values created alignment. The project forged ahead with stated values but still devoid of power-shifting strategies. The continued lack of a power shift strategy created the conditions for an inevitable break in tenuous trust. For our project, trust was broken when the white male contractor used the learning from the project gleaned from Black and Indigenous leaders to promote his own business, promoting the white male's newfound racial equity insights without getting their permission or crediting their contributions. This action is an outcome of power. It is privilege to assume that knowledge gleaned in a collective setting can be used outside the group without the explicit permission of those from whose intellectual property a new insight was developed. This action is also indicative of a pattern of the extraction of resources (physical, emotional, intellectual) that white supremacy culture normalizes, which then triggers the reactivity of whiteness when this pattern is called out and explicitly named.

The use of group intellectual property without group permission and proper citation was a clear violation of the guiding principles established during the team's racial equity training series. When the team learned of this breach of trust, the white male contractor was confronted and summarily denied the breach, reflective of the normalization of this action in white supremacy culture. The Black and Indigenous leaders in the group named the harm, were explicit about why it was a problem, and named the impact on their lives and livelihoods.

This cycle of harm is specifically dangerous for the Black and Indigenous leaders, with Black and Indigenous women bearing a disproportionate burden of this harm. As an Indigenous woman, Vivette expounded, "These conversations can cost us. Sometimes our lives, our family's lives. It can cost us economically with jobs or consulting work. It can even cost us in relationships. There's a cost to standing up. You're putting us in harm's way for speaking up because we made a white man uncomfortable."

There is a responsibility in this space for white people, which is often not taken; as Gretchen recalled, "I felt a sense of alarm, and I also remember not saying anything at that moment. As a white woman I *needed* to step up and call this man out directly, but I retreated into complicit silence." Gretchen didn't want to interject at that moment because she didn't want to be paternalistic, like the white savior needing to step in. But this focus is misplaced on making sure that she did not look bad rather than calling out harmful language that perpetuates injustice. Also, Gretchen wanted to keep the peace and not disrupt the group, and there are clear norms within white culture to not call out or embarrass white men. When coupled with the power differential between the white man and the group, these normative understandings and the internalization of these norms made a stance of complicit-ness the safe retreat for her in this space.

Typically, the cycle of harm continues as PWIs maintain focus on the deliverables and exit the partnership when conflict is called out. The commitment for the PWI is achieving the outcome of the creating the plan on time and within budget. The Black and Indigenous leaders in this project expressed that this has been their repeated experience. Vivette shared, "When you allow white folks into your community, there's always a risk that the next thing you know your culture, your food, your way of being here in your language will be commodified and sold back to you. When you say they come in and bring justice, they're bringing white justice, right? That is the savior complex to them. It allows them to remain comfortable because conversations are avoided. Then I am uncomfortable because avoidance is counterintuitive to how I experience the world."

Black and Indigenous people involved in PWI-led food system work repeatedly experience this cycle of harm. Typically, this is where PWIs exit, denying that any real harm has been done because the process produced an output, whether that be a plan, a paper, or simply convening the group. As Justin noted, "How you start is where you typically end."

Refusing to Commodify Justice

This is where our story diverged from the common narrative. Ending a project in a process of harm is not inevitable, but there is no institutional model or incentive from PWIs to do things differently. Vivette shared, "We haven't lived in a system that's different than this. But we can. We can create something brand new." For our project, the first step in breaking the cycle was stopping when the harm was inflicted. We made an active decision to not push ahead. Stopping this process was not easy. It required cover, relationship, and accountability.

As a white woman working in PWIs, Gretchen explained, "Stopping a process within an organization that depends on project deliverables and timelines, is like trying to stop an airplane carrier. Nearly all the momentum in PWIs is geared towards bringing in more money, building a portfolio, and chasing that next deliverable. There isn't any room in these systems for doing equity and justice work, so it can result in blowback." Equity work is about relationships and trust; it takes time, has no visible product that can be returned to the institution, and has no model or roadmap to achieve. Institutionally, the focus is timelines and deliverables, so the ability to attend to relational and emergent needs is undermined by the requirement to produce a product. This perpetuates alienation between institutions and communities, where institutions serve at the altar of production and at the expense of each other.

Cover was needed at a leadership level for the work to pivot from a financial-, output-, and time-driven process to a people-centered process. The white male academic who initiated this idea created critically important space, using his power to let relationships evolve. Jen recalled, "It was the director of our center at the time that used his power to cover me within the institution to NOT produce a plan. While, like me, he didn't understand the nuance of everything that was happening, he cared about the human beings involved and sought, first and foremost, to protect those relationships."

At this juncture, cover was also required for the PWI project evaluators. This meant that the design team members, now both BIPOC leaders and PWI representatives working together, needed time, the removed burden of the need to produce a product, and the freedom of a safe space where relationships could heal and evolve. Moreover, this process needed to be protected from potential negative repercussions from those within PWIs who may not agree or may take issue with the process. Natalie recalled, "To write about observations of centering whiteness, harms, and decolonizing approaches . . . authoring those words could potentially have repercussions for me as a Black woman that were different than my more senior level white colleague. However, she covered me and stood alongside me when reporting on our work internally."

Protecting relationships required space and time. The group needed to address the harm that had impacted the team. Camryn named the role of white people in this space, "She [Jen] blocked for us [prevented others from harming] and protected us, and she trusted us. It wasn't just because we were Black and Indigenous women. It was because she knew what we were saying [about how to be in relationship, how to de-center whiteness, center BIPOC leaders, and truly transform the food system through power shift and reparative strategies] was true and that the problem was much deeper."

If we were to authentically engage in a decolonized food justice strategy, we had to make the move that PWIs do not typically make—move from transactional to relational and humanize what was happening, acknowledging that systems are made up of people. Vivette emphasized, "We all talk about the system as if the system is separate from us, or somehow, we're not part of this system. The way the system is set up is either you're bad or you're good, we've got to humanize this system. Look at the faces, name those faces. Name the people."

Humanizing the system also means deconstructing internalized, white-centered norms. That requires relationship with oneself and accountability to one another, as opposed to accountability to money, timelines, and outputs. Camryn noted, "How Jen, being as young as she was in her racial equity analysis formulated enough relationship with me, and enough care for me that she blocked for us and protected us. That was because she knew that the only way over and to be successful was through, and that we were the ones that would guide it."

By humanizing the system and moving in relationship and accountability to one another, this story moved from a cycle of harm into a cycle of healing. Vivette described the painful healing process: "You've got to go through and name the harm and the mechanisms of harm. It is about not blaming. It's naming the system that we're operating in. We can't dismantle what we don't name." She added, "Growth is uncomfortable."

In a typical PWI process, when harms are inflicted, they are minimized or addressed behind closed doors, recentering whiteness and increasing harm to Black and Indigenous leaders. Camryn noted her experience: "Because of how whiteness operates, we flip the blame to Black and Brown people for trying to hold white people accountable. It's decolonizing, holding people accountable because white systems and structures will not do it. You need to understand that systems that white people have set up have this shared understanding about gossiping, but it's a whole other thing if you're trying to hold someone accountable and mitigate the harm that they cause, especially when they're in a position of power."

The group brought in a mediator to address the breach of trust by the white male contractor. The mediator brought the full group into the room where the project planning had been underway to walk through the impacts of the events on the people participating in the work. Justin expounded on the distinct difference our process took by stopping and creating a space to name and address the harm with the goal of healing: "As a Black man in white spaces, we don't get to have that experience very often [conflict and resolution] . . .

it's just usually very disjointed. It feels very important to recreate the situation literally . . . the place in which it happened, how it was going on, what was it that people were experiencing or not experiencing so that, you can really be able to close the book on this. It felt like the way that conflict could truly be resolved.

As a result of the mediation, the white consultant was fired from the project. Jen recounted, "I don't think there had ever been repercussions for his bad actions before, because he could always get absolved by claiming good intent. We protect intent in PWIs more than we protect the Black and Brown people that 'intent' impacts." Following this change in group structure, Justin named the change in our interactions, recalling,

> After that process, one of the other members of the group was just way more talkative than he had been, and somebody mentioned that. And he was like, yeah, I didn't trust that white man, and then everybody laughed. Everybody laughed for like five minutes straight. It was almost like a TV show comedy moment laugh, right? It felt like one of those cleansing, rinsing laughs. We had to experience this all together, and it felt like the moment of levity that broke everything down. It was like we got to experience the harm collectively, but we also got to experience the healing collectively.

This key moment highlights the dilemma faced by Black and Indigenous communities doing this work in partnership with PWIs. When they are forthcoming, holding space, naming harm, demanding accountability; they are labeled as difficult, discredited, and sometimes removed from the process while their ideas are co-opted. When they preserve their ideas and safety within the space, they are not able to share and lead authentically at the full capacity that they possess, which in turn PWIs can label as disengaged or ill equipped.

The team members working in PWIs faced different dilemmas. Institutional norms do not support power shift strategies. When white team members maintained the process in line with the project objectives—convening a group that included Black and Indigenous leaders, meeting with the intention to create a plan within budget and on time—they received full support from their institutions and other PWIs. When they slowed the process, questioned the intent, made changes to compensate and center Black and Indigenous leaders, they encountered pushback within their organizations that required cover from supervisors in positions of power. Pushing back on institutional protocol creates the risk of losing reputation within the organization, removal from the project, or potentially losing one's job. Navigating these risks within PWIs required the white women in this process to develop a stronger analysis of power and the impact of privilege and to develop skills to move against the norms of their organizations. The white women in this project had to make an active decision to, as Camryn stated, "be who you say you are" and choose people over process.

For this project, the result of stopping and creating a collective space to name and address harms moved us from a cycle of harm to a cycle of healing. While more harm was inevitable due to the reinforcing systems and structures of PWIs that will continually reward alignment with normative PWI protocol and deter power-shifting strategies, there became a new way of being.

Discussion

"We can't do decolonized work when the whole thing is shaped from oppression," shared Kamal. A transactional, product-driven process initiated by a PWI will never result in a more equitable and just food system. It must be a process that humanizes the people in the system, taking time to earn and build the trust and relationship that is necessary so that inevitable harm does not destroy the effort. Justin explained, "Community does not necessarily mean harmony. Community means conflict. Community means conflict resolution. All those things are all in there together."

A decolonized process sets a way of being that acknowledges that harm will happen, creating spaces to address harm and begin to heal. Vivette offers the four R's that are the core elements of a decolonized way of being: "responsibility, reciprocity, right relationship and redistribution" (Harris and Wasilewski 2004). She shared,

> Because, if I don't have, I am clear that I'll be taken care of. Regardless of what I have, if I know someone else doesn't have, it is my responsibility to share with them. [It is about] giving away, not amassing. It's the opposite of monopoly, which is colonization. The riches are about the relationships. Reciprocity is different, and it's not transactional. The reciprocal is like, I'm giving freely without any understanding or expectation that you will pay me back because I'm in relationship. And, when I'm in need I will be taken care of by someone other than the person that I shared with knowing it's being in community. It's circular.

To understand repair, we must acknowledge the US history of accumulated harm and trauma. Addressing that history is a start, but it is not a fix.

"You can't put a timeline on healing," shared Kamal. The work will be slow. It will not fit a grant or project timeline. It must be done in relationship. Vivette emphasized, "None of us can singularly do the undoing. It takes all of us. Which is also a part of decolonization."

Decolonized work requires redistribution—a shift in power. As Vivette noted, "It is first turning the mirror to yourself to understand how each of us are perpetuating the system, and then, seeing what redistribution or repair can be brought to bear." Why is it so rare for a PWI-initiated decolonized process to happen? The policies and procedures of PWIs are designed to protect the PWI. The institution operates in a system of rewards and reinforcements that incentivize actors within the institution based on meeting the standards of the system. "I could be very successful in my career and never engage in a decolonized process." shared Jen. "There is no policy in my institution that rewards my actions of slowing things down and building relationships." Kamal added, "When we're talking about decolonizing work, we must change our relationship with finances and money as the motivator and basis of the work. How does a system look which isn't predicated on revenue, but actually getting food to people who need it?"

In PWI processes, funding drives everything. It defines a grant cycle, fiscal year, percent of effort, scope of work, deliverables; all actions are tied to time and money. Decolonized work is lifetime work, which does not align with PWI timelines. Gizem recalled, "When the project came to a halt and it shifted, there was an obvious duality between the PWI demands and the emergent process within the group of community leaders. It was mind boggling to code-switch between these two worlds, even though we are talking about the same process."

Additionally, there is risk within PWIs to counter its system of rewards and incentives. A decolonized process requires actors within PWIs to push back against timelines and predetermined objectives and outputs, which may result in organizational change but more often results in rejection and retrenchment. Gizem expounded on her experience as an immigrant woman in PWI spaces: "Traditionally in PWIs, when you name a problem, it is either swept under the rug, or the person who names it becomes a target of unjust scrutiny."

Retrenchment comes in many forms, such as terminating the work, terminating the person/people, and/or rejecting the narrative. The retrenchment impacts lives.

What Was Different About This Work: We Didn't Solve Everything, but We Did Move Differently

What was different about our story is that how we started is not where we ended. We experienced the crucial components of decentering whiteness, humanizing the system, creating shared values and ways of being that allowed for collective healing, and staying committed to the long work of individual reflection and relationship building. These components must occur for any social change, including food system change, to take place.

We are not offering a solution for other communities. For us, moving from a traditional process to a decolonized process was the outcome. Justin reflected, "A decolonization approach must be a part of these kinds of conversations because, without this kind of approach, I think we would have just ended up writing some [expletive], and everybody would have gone home disgruntled. [A decolonized approach] stopped the process from going fully all the way left."

There were specific outputs as part of this process. We produced products that answered questions raised by the team. "Power and Benefit on the Plate, a History of Food in [our city]" was a means of identifying the history of policy and practice that created the inequities in our food system. This narrative laid the groundwork for future policy efforts, rooted in a racialized historical analysis. We guided a student thesis to create a research brief, "Identifying and Countering White Supremacy Culture in Food Systems" as a means of naming what is typically done in food security work, so that we can move differently. Prior to the disruption of this effort by the COVID-19 pandemic, the group hosted caucused dinners to build broader relationships over food, to offer the healing space and the visioning space of what a decolonized food system in our city might be.

The outputs, though, are not the result. The result of our work together is the process itself, which unfolded as our story. If we want food justice, we need to learn from our stories.

As Vivette shared,

> This is how we must move, and it's not a recipe. . . . It's just a story . . . you can't take this and replicate it. It's like, here it is. This is how I learned at the feet of my elders. How I learn—they tell a story. . . . And that for us, one critically important aspect of our story is, the emergent process of harm and healing. . . . Because that's not usually lifted in articles or stories about this type of work. And I think

> it's powerful because it also can serve as knowledge and foundation . . . it's normal to go through this when you're doing the decolonization work.

For our story, we offer the process. The process of stopping, humanizing, decentering whiteness, and creating conditions for decolonized work. It is lifetime work, not timeline work. For those PWIs committed to justice, it is simply the work.

Note

1. Community-rooted™ is the trademarked approach to community development created and implemented by Communities in Partnership in Durham, NC. The trademark has been essential to protect the intellectual property of the majority-Black women who developed this framework.

References

Bourke, Brian. 2016. "Meaning and Implications of Being Labelled a Predominantly White Institution." *College and University* 91 (3): 12–18, 20–21.

Collins, Patricia Hill. 2022. *Black Feminist Thought, 30th Anniversary Edition: Knowledge, Consciousness, and the Politics of Empowerment*. Routledge.

Estable, Alma, Mechthild Meyer, Canadian Labour Congress, and Gordon Pon. 1997. *Teach Me to Thunder: A Training Manual for Anti-Racism Trainers*. Canadian Labour Congress.

Harris, La Donna, and Jacqueline Wasilewski. 2004. "Indigeneity, an Alternative Worldview: Four R's (Relationship, Responsibility, Reciprocity, Redistribution) vs. Two P's (Power and Profit). Sharing the Journey towards Conscious Evolution." *Systems Research and Behavioral Science* 21 (5): 489–503. https://doi.org/10.1002/sres.631.

Hummel, Charles E. 2009. *Freedom from Tyranny of the Urgent*. InterVarsity Press.

National Museum of African American History and Culture. n.d. "'Whiteness.' Talking about Race, Definition of Whiteness." Accessed June 29, 2023. https://nmaahc.si.edu/learn/talking-about-race/topics/whiteness.

Racial Equity Tools, Glossary. n.d. "Definition of Decolonization." Racial Equity Tools. Accessed June 29, 2023. https://www.racialequitytools.org/glossary.

Smith, Camryn, Danielle Spurlock, Aliyah Abdur-Rahman, and Kay Jowers. 2020. "Community-Rooted Organizations: Enhanced Accountability and Capacity Building for Community Development." *Métropolitiques*, July. https://metropolitics.org/https:/metropolitics.org/Community-Rooted-Organizations-Enhanced-Accountability-and-Capacity-Building.html. Accessed June 29, 2023.

14 White Masculinist Pasts→Black Feminist Futures: Lessons from Spelman College's Victory Garden as Black World-Making

Whitney Barr and Nik Heynen

Tucked behind the speckled shadows of Giles Hall on the campus of Spelman College in Atlanta, Georgia, is perhaps the most biodiverse and regenerative space on an American college campus that unapologetically centers Black women, African diasporic foodways, and living relationships with health and food, land, and agriculture. This historic landscape is referred to as the victory garden on campus because of its previous use by and for students, faculty, and staff during and after World War I. At various times since Spelman moved to its current campus in 1883, food has been grown across campus in the form of dairy farming and agricultural crop production. The present-day victory garden serves as a living historical marker of this legacy while also helping us envision just and sustainable food futures. Originally, the land was occupied by the Muscogee Creek Nation and, before becoming the home of Spelman College, it was used as the Fort McPherson military base. At some point after the world wars, Spelman's victory garden fell out of use. And then, contrary to the large-scale, white, masculinist imagination behind the nationwide victory garden movement, beginning in 2009, a group of Spelman faculty began to revive the garden by growing food on-site again. In 2014, Spelman's Bonner Office of Civic Engagement cultivated a service-learning initiative entitled Sisters of the Soil in partnership with HABESHA, Inc., a local food justice organization. This initiative grounded students in relationship with urban agriculture on campus and throughout the Black West End community. This historical arc contributed to Spelman launching the first interdisciplinary food studies program at a historically Black college or university (HBCU) in 2015. Through this program's garden, we reimagine what victory means in the context of Black feminist food futures. As a deliberate response to the history of the victory garden movement, throughout

Figure 14.1
Students working in the Spelman victory garden in the 1890s.

the remainder of this chapter, we will refer to this living landscape as "the garden," though its formal name has not yet been changed.[1]

Figure 14.1 dates to the 1890s and depicts seven Spelman students tending the garden outside Giles Hall. It offers incredible historical insight into the importance of studying and understanding food and land for Black women at Spelman over 130 years. Speaking about this photo, former Spelman food studies codirector Ashanté Reese (2018, 1) suggests, "Clad in long skirts, armed with farming tools, the unidentified women grew food as part of their education well before food studies had a name or structure." According to Reece, Spelman's food studies program trains curious, theoretically engaged, justice-focused food scholars and activists (Reece 2018, 5). The garden now serves as a physical and living anchor of this growing academic program, and the authors of this chapter have close ties to the garden. Whitney, a Black woman, landscape designer, and Spelman alum, served as the program and garden manager from 2021 to 2023. In that role, she recruited Nik, a white male activist-scholar from the University of Georgia, to serve as a visiting scholar and teacher within the food studies program.

We developed our relationship through years of supporting Saltwater Geechee descendants from Sapelo Island, Georgia, in their efforts to make Geechee foodways more physically visible and protected. Insights we gleaned together on Sapelo Island are central to our thinking about the garden.[2] In short, Gullah Geechee people descended from West Africans were formerly enslaved on island plantations because of their agricultural expertise. Nik worked closely with Ms. Cornelia Walker Bailey, a Sapelo Island descendant, matriarch, and griot, and her memoir inspired Whitney to live "on-island" and to retrace and uplift Bailey's detailed medicinal plant memories through a reflexive practice of racial healing, landscape design, and crop cultivation. She inspired us both with her vision of resisting plantation-derived Black erasure through Black land ownership and everyday land practices such as growing Saltwater Geechee red peas and other heritage crops.

Through our work with Ms. Bailey and other descendants on-island, we began to think with the notion of Black world-making to understand how garden-based practices at Spelman create everyday utopias (White 2018) and resist commonly accepted themes found in plantation landscapes, including the exploitation of land and the Black woman's body, production, and Black women's agrarian erasure. Getachew (2019) puts forth world-making as building a more just world in the context of global efforts to fight against colonial domination and white supremacist oppression. This forward-looking framing offers us the ability to think about how local actions can speak globally. Táíwò (2022) builds on Getachew by framing world-making as not just focusing on the redistribution of wealth and resources but also making and sustaining environments that facilitate everybody to grow, heal, and flourish. The historical and ongoing significance of the garden and the young Black women who offer their stewardship, attention, and learning within it, comprises an embodied practice of world-making through a regenerative, African diasporic planting approach that centers Black women's imagination, creativity, care, and power. In an intentional move to make the concept of world-making more concrete through diasporic landscapes, Spelman students have also participated in on-island land practices in tandem with their garden experiences on campus. Within this small, layered garden, Spelman students hold spatial agency over their food futures and a tactile *choice to change the world.*

This world-making work is necessary because the plantation is still with us, and we live, gather, and grow food in the wake of its afterlives (Hartman 2007; McKittrick 2013; Sharpe 2016). For many Black Americans, food and agriculture are sites of historic and ongoing racial trauma, from the sale of labor and personhood to the creation of contemporary food apartheid (see chapter 6) to the demonization and mocking of some of our food traditions (Williams-Forson 2022). In this context, Spelman's victory garden works toward racial healing by reimagining the future of Black foodways and Black women within food studies scholarship. Black world-making helps us imagine space for racial healing and Black women's empowerment at the scale of specific Black foods as well as the colonial and white logics that appropriated them as Southern and American, as opposed to Black (Twitty 2017).

This chapter also situates Spelman College's victory garden, an intersectional learning greenspace grown by, with, and for Black women, as a model for reestablishing and developing future campus gardens—particularly throughout HBCUs. The garden recenters African diasporic foodways and practices while empowering young Black women to experientially address food system challenges and contradictions. We are interested in situating this growth and reimagining of what "victory" means today. Counter to "victory" as relating to the white, masculinist, capitalist, and industrial agricultural definition invoked during WWI, we are interested in redefining "victory" to support liberation and Black world-making for, by, and with Black women. Spelman's garden prioritizes agrifood learning, rest, sisterhood, and reciprocity with land. The garden is also a living intervention model in response to enviro-racial traumas experienced in Atlanta and throughout the world.

Whose Victory?

Multimillionaire and forestry expert Charles Pack initially proposed the idea of "victory" gardens in a 1919 report through the US National War Garden Commission while he served as its president. Arguing that "the war garden must become the victory garden" (1919, 4), Peck announced the transformation of what had been a large-scale effort during the First World War into gardens through which people across the US would produce and ship large volumes of food to Europe as countries rebuilt. Pack's approach built upon what had been called "war gardens" during WWI by updating previous propaganda to enlist "patriotic" people across the US to begin producing large amounts of food across their personal yard space and other vacant lands.

Pack started his report by stoking patriotic action through a vision of racial capitalist growth and erasure (1919, 9): "No other single business born of the war has affected a greater number of people than has gardening. Starting from a mere nothing before the U.S. entered the war, this form of service grew in less than two years into a new occupation, which counted its followers by the millions and, in the number of people employed, exceeded any other branch of gainful occupation, with the single exception of actual farming." Importantly, in 1920, when Pack was making his stirring plea to the white gaze central to the era's agrarian imaginary, 560,000 people, mostly Black, were working as sharecroppers across the agricultural south, a system of farming that extended brutal plantation logics well into emancipation (Alston and Kauffman 1997).

In the current moment of increasing conversations around land, racial justice, and equitable food system futures, the whitewashed victory garden movement and Pack's extractive vision of "putting all idle land to work" (1919, 10) requires critical attention. Black veterans and Black farmers face discrimination, then and now, but this history has too often been glossed over with energetic, white feminist-centered victory garden marketing propaganda *sans* Black women. We could not find one victory garden poster (there were many) that represented Black people or Black life. Recently, the Black Lives Matter movement brought worldwide visibility to the policing and discrimination of Black people in public spaces and the lack of Black land ownership (for agency of space) and reignited interest in African diasporic land and community practices, which have largely been cultivated by Black women (see Penniman 2018, 5). Concurrently, COVID-induced food anxieties helped resurface an interest in victory gardening at home and helped pique the Spelman community's interest in the garden's 2021 revitalization. But the ways in which the garden is cultivated and cared for, beyond production, makes it more than just a finite victory. As part of this ongoing movement, the garden is an educational Black food geography that unapologetically reshapes the whiteness of the victory garden mentality by centering the needs and interests of Black women on campus.

Spelman's (~~Victory~~) Garden as Black World-Making

Spelman College, a liberal arts institution designed for Black women, has been ranked as the #1 HBCU by *U.S. News and World Report* for the past seventeen consecutive years. Spelman has served as an important hub of

Black feminist thought and activism, which is especially significant given that agricultural education, geography, and home economics were some of the earliest programs within the curriculum (Lefever 2005).

In 2021, after a period of dormancy during peak pandemic years, Whitney led the revival and iterative design of this sacred organic space, alongside like-minded students, faculty, staff, and alumna, into a site of Black (green) refuge deserving of a new name—beyond its oversimplified and military-focused label of "victory." When Whitney returned to her alma mater in 2021 as the program and garden manager of Spelman's growing Food Studies Program, she quickly realized that the garden helps facilitate a healing process related to racial traumas connected to the history of US agriculture. The space also, in practical terms, helps to ground Spelman's Food Studies Program and to nudge critical dialogue concerning Black agrarianism and Black women into collaborative action in alignment with Black (eco)feminist futures (McCutcheon 2019).

We have both witnessed the power of Spelman's diverse student population and how the garden's diasporic plant material of over eighty different varieties has something that resonates with each of us. Caribbean-descended students from Brooklyn find Trinidadian pigeon peas, while Nigerian women find vibrant *efo shoko* growing alongside ultracross heirloom collard greens. Beginning in 2021, students learned how planting Palestinian *molokhia* within the garden and saving its seeds is a localized, world-making practice of resisting cultural erasure. Some have been inspired by the ADA accessibility of modular metal beds punctuated throughout the garden to casually stroll and take interest in understanding how interplanting calendula with brassicas may support pest management and phytoremediation. Some students have gravitated toward the garden as a moment of relief and engaged in sensory-motor practices of touching hanging hibiscus-based paper or smelling the differences between lemongrass, lemon verbena, and lemon balm. Student volunteers have often been in awe of being able to create garden-based fertilizers. Some have engaged in the contemplative experience of discovering what their Black bodies may need or crave amid the garden's thoughtful ecosystems. One involved student volunteer designed 3-D-printed tools that make sowing small benne seeds easier after remembering how long it took them to thin out seedlings with their Spelman sisters.

The garden has also garnered cocurricular involvement from Spelman faculty. One class engaged students in dance activism[3] through a bodily response to forced labor, regenerative labor, and discovered liberation within the garden. The openness of the garden's landscape prompted students to reflect on the luxuries of Black picnic rituals and mindful eating with community while paying homage to the land and ancestors. Some students enjoy dancing while scattering native pollinator seeds for the land's needs and pondering graduate school, professional, and entrepreneurial opportunities that can extend these regenerative moments in the form of food system research, policy, advocacy, technology, business development, and art. Though small, the experiential space the garden provides is immense.

When prompted to discuss her main takeaways in a garden log, one student wrote, "[I learned that] some seasons are for *reaping* and others for *sowing* some things for *harvesting* others for *plowing*. Most seasons are for just *growing and learning*. Expecting the same thing season after season is a shortcut to being disappointed. But embodying ourselves in what each season has to bring is only a pathway to greater experiences and knowledge." She went on to write,

> One of my major takeaways this year was about solidarity, understanding who we are even in solidarity with and how we can be a more solidarity minded community. I think the garden is central to this. When I think about opening access to food equality throughout the AUC [Atlanta University Center] and Atlanta in the south in general, I think about making sure we're buying ethically sourced things or empowering communities who create crops to keep themselves alive. That solidarity doesn't come without actual work. The literature tells us that we need to be an actual community with others. We shouldn't focus on a larger systematic downfall but towards a prefigurative community structure.

By prioritizing moments of collective cultural memory and Black women-led land preservation, Black foodways can stay alive, and we can avoid a flattened landscape of repetitive, limited crops that are not resilient to climate change. The moments when students ask to have a photo taken to capture their collard green harvest and share with family back home or when students seek out red okra for a women's health steam—these small moments help stitch together seemingly invisible Black food geographies by simply existing, triggering nostalgia, and building new connections, experiences, and emotions related to agriculture. These experiences aid healing (see Hassberg 2020; Garth 2020; chapter 4 this volume).

While Black folks, and especially Black women, were excluded from the vision of the federal Victory Garden Project, the garden provides an important moment for Black world-making, which is necessary for racial healing across antiracist edible landscapes. The garden crafts space for casual, nourishing encounters as well as a shared language and plant pallet for antiracist food futures. This seems particularly relevant on college campuses where students are encouraged to develop skills needed for transformative careers that address the complex challenges and opportunities of our global food system.

Food Studies as Curriculum and as Racial Healing

We worked together to nurture our understanding of Black world-making through the garden and a class we cotaught called Food in the City. We agreed to teach the class together so we could iteratively test things out, feel them, make mistakes, laugh, find joy, recharge, and find reasons to return to the regenerative work both in the classroom and in the garden. We think it is naive to assume that students will just stumble into food system jobs. Spelman's garden design aims to prepare students for antiracist career roles in *food futures* beyond existing food markets. Academic institutions invest resources toward traditional employment opportunities, and we see value in that, but we see the experiential learning within the garden as something much more holistic and timely. Thus, the garden can be positioned as an encounter between more typical university priorities and the world-making work of land-based learning.

During the Food in the City class, we talked about how important it is to stop using Black people as exploited sites of production and labor for traditional agrifood spaces. We focused on a spirit of building from existing cultural and physical assets through the garden and the classroom. Within the garden, there is an ongoing academic process of reimagined relationships between Black women, food, and land. And because the garden prioritizes diasporic food systems rather than reducing itself to the sole victory of extractive agricultural yields, it functions as a Black food geography that is successfully destabilizing place, space, and Black invisibility (see McKittrick 2006)—all of which are needed in the design of equitable and just food system futures. This is not a space that requires one to decide between land as a regenerative spiritual place or an extractive site solely for wealth building; we see both because we are complex beings and deserving of it all.

As we created the curriculum together to reflect these sentiments, Whitney wanted us to sit with the notion that "the process is the point." For example, the process of collectively learning to grow foods superseded the end product "victory" of yielding large harvests.

Spelman's Food Studies Program has emphasized small, deliberate steps, small classrooms, small outdoor growing labs, and grassroots organization within the ironically hierarchal gates of an academic institution embedded within the food apartheid of the historic West End Atlanta community. We wanted to focus on nurturing an agroecological, climate-resistant, and unapologetically Black space and codesigning *with* rather than controlling the landscape *for* human needs only.

Insights gained from our discussions with students suggest that we need more free-form spaces where mistakes can be made, collective garden values can be established, and students can be themselves—absolved of any assumed green standards of whiteness or perfection (DuPuis and Goodman 2005). This has been pointedly observed when students have asked about the use of specific plants for Black women's physical and mental well-being and how they can get involved even when they think that they have no green thumb or can't cook or can't (fill in the blank). Typically, students connected their introduction to food and agriculture to a woman in their family. Throughout their experiential, garden-based education, they are reminded that they can do all things and that there is abundance within small spaces that are cared for—food spaces that Black women have always helped lead and sustain in this country. Spelman student garden volunteers also engaged in lessons about activity below the soil's surface that influence the agroecological performance of perennial food systems. For example, when the low temperatures of winter 2022 created voids above southern grounds, including the garden, students questioned the impacts of their ongoing labor in seemingly "dead" food spaces. In response, we worked with the soil and redirected collective class energy to relocating salvageable plants indoors, resowing medicinal herb seeds, and sipping Spelman blue butterfly tea over class discussions.

Across our experience, we saw students imagine and connect the garden's role in prepping for food careers and centering mindfulness throughout the healing process. Students described feeling more and more empowered to pursue food-entrepreneurial opportunities through value-added consumer goods such as medicinal teas and mushroom-infused soaps. But in contrast to production-focused agriculture, the garden also leaves room for Black women to physically gather within the space and dream up these food futures together, which they did (and do) regularly. Faculty- and student-led groups also have found ways to convene within the garden and to rethink decolonized food systems through cocurricular coursework, service learning, and team-bonding exercises.

Through our teaching together, many things have become clear. Foremost among them is that Spelman women tap into their intersectional connections with food and land and their daily campus life experiences related to food, space, and agency. They are asking critical questions of our food system's complex and siloed networks of injustice. They are actively engaging in agrifood studies through adrienne marie brown's (2017) "small is all" approach with the bold thoughtfulness needed for food system reform. For years, Spelman students have advocated for local food procurement within the cafeteria. Resurfaced garden momentum helped lead to the college's decision to begin sourcing from local Black farmers. This move, nudged by the physical garden, affirmed student-led food justice and climate-focused activism.

They are also working through a prefigurative "everyday utopias" mindset as discussed by White (2018) to imagine new food futures. Their activity within the garden, food studies classrooms, and everywhere they go reflects their willingness to shift the rules of (food) power.

We are growing together not only at a pace that reflects seasonality and strengthened relationships with the land, but at the pace of authentic bonds being built among Spelman food scholars. Within the garden, students are liberated from traditional classroom expectations and encouraged to lean into sisterhood and Black women agrarianism. In that same space, we have also seen the ethos of white masculinity present itself when students timidly engage with the garden—instantly questioning what value they can offer to the space or doubting their innate connections to land/food. And we have seen the shift in students from being nervous, lone volunteers looking for a place to "fit" in this "post"-COVID environment to Spelman sisters blending up herbal tea blends while singing along to Beyoncé's *Renaissance* album during a Food in the City course. Motivated by Fannie Lou Hamer and the garden's fruition, students proudly harvest produce for commuter students battling food insecurity. They are growing together because they want to and not because they have to. Spelman's legacy of valued sisterhood already binds them across campus, but the interactions, no matter how small, within this special, yet hidden garden just seem to offer a different energy to the campus landscape that surpasses the definition of a victory garden.

Conclusion

As we conclude this chapter, it is difficult to not ponder that SCOTUS recently ruled to end affirmative action in college admissions. While it is too early to know how this decision will affect HBCUs, our experience suggests that now more than ever, we need to preserve, revive, and reestablish Black edible landscapes such as the garden at Spelman. In line with Garth (2020, 109), we want to argue that Black spaces do not need white-adjacent organizations to affirm their value of existence or to lead more instances of utilizing Blackness to legitimize its higher education food justice work while at the same time policing the way in which students express their Blackness and engage with Black food culture. With Black spaces being threatened by the ending of affirmative action, the increasing erasure of Black-owned land and Black women's key involvement, we have to hold on to, preserve, dream up, grow, and steward land for Black feminist food futures, as this is work of Black world-making; making a new world that includes all of us.

The intimate one-eighth-acre garden offers a great opportunity to revive and reimagine agroecological relationships and Black life. As we are seeing at Spelman, HBCU campus gardens play a distinct role in inspiring Black land ownership, encouraging intergenerational family foodway preservation, shifting agrifood perspectives, and centering Blackness in the garden. These are the processes of Black world-making.

Black world-making becomes visible when we go out to the garden with students to harvest medicinal herbs and witness their reconnection to plants that their mothers and grandmothers helped them first conceive of. It is a gift and a spiritual journey we have witnessed in line with former Spelman student Alice Walker (2004), who has written about the journey to be *In Search of Our Mothers Gardens* (2004). For us, Black world-making is seeing students lean into self-sustenance, Black joy, and sisterhood in an on-campus green space in community with other resilient, diasporic landscapes; Black world-making means to be in relation with local growers, herbalists, stewards, and other critical food system leaders. If this type of world is possible on one liberal arts campus in Atlanta, we wonder how Black world-making can be cultivated across the college landscape, particularly at HBCUs, to advance food studies scholarship, to decolonize garden practices, and to help drive future food system careers.

Notes

1. It is necessary that we uplift Sariyah Benoit, who agreed to fervently work alongside volunteers and food scholars from 2021 to 2024 in order to activate, nurture, and maintain the garden.

2. We initially came to think together about Spelman's garden through shared commitments to making Gullah Geechee foodways on Sapelo Island, Georgia, more physically visible and connected to Black Southern foodways. Working and learning together for several years on Sapelo Island was pivotal in our collective efforts to forge solidarities with each other and a group of Saltwater Geechee descendants on Sapelo Island, primarily Ms. Cornelia Walker Bailey and her son Maurice. Several years of working together on Sapelo Island inspired and gave greater meaning to a new collaboration through Spelman College's Food Studies Program when Whitney became the program and garden manager in 2021. While managing the program, Whitney helped recruit Nik to start teaching in the program as a visiting scholar.

3. Dr. Julie Johnson's Dance, Community, and Social Change course.

References

Alston, Lee J., and Kyle D. Kauffman. "Agricultural chutes and ladders: New estimates of sharecroppers and 'true tenants' in the South, 1900–1920." *Journal of Economic History* 57, no. 2 (1997): 464–475.

Barr, Whitney. *Designing for Racial Healing: (How) Can Physical Design Responses to Plantation Futures Offers Space for Racial Healing?* Master's thesis. University of Georgia, 2021.

brown, adrienne maree. *Emergent Strategy: Shaping Change, Changing worlds*. Chico, CA: AK Press, 2017.

DuPuis, E. Melanie., and David Goodman. "Should We Go 'Home' to Eat? Toward a Reflexive Politics of Localism." *Journal of Rural Studies* 21, no. 3 (2005): 359–371.

Garth, Hanna. "Blackness and 'justice' in the Los Angeles food justice movement." In *Black Food Matters: Racial Justice in the Wake of Food Justice*, 107–130. Minneapolis, Minnesota University Press, 2020.

Getachew, Adom. *Worldmaking After Empire: The Rise and Fall of Self-Determination.* Princeton NJ: Princeton University Press, 2019.

Hassberg, Analena Hope. "Nurturing the revolution: The Black Panther Party and the early seeds of the food justice movement." In *Black Food Matters: Racial Justice in the Wake of Food Justice*, edited by Hanna Garth and Ashanté M. Reese, 82–106. Minneapolis: Minnesota University Press, 2020.

Lefever, Harry G. *Undaunted by the Fight: Spelman College and the Civil Rights Movement, 1957/1967*. Macon, GA: Mercer University Press, 2005.

McCutcheon, Priscilla. "Fannie Lou Hamer's freedom farms and Black agrarian Geographies." *Antipode* 51, no. 1 (2019): 207–224.

McKittrick, K. *Demonic Grounds: Black Women and the Cartographies of Struggle*. Minneapolis: University of Minnesota Press, 2006.

Penniman, L. *Farming While Black: Soul Fire Farm's Practical Guide to Liberation on the Land*. White River Junction, VT: Chelsea Green, 2018.

Reese, A. *What It Means to Study Food at an HBCU*. Southern Foodways Alliance, March 29, 2018. https://www.southernfoodways.org/what-it-means-to-study-food-at-an-hbcu/.

Táíwò, Olúfẹ́mi O. *Reconsidering Reparations*. Oxford, UK: Oxford of University Press, 2022.

Twitty, Michael W. *The Cooking Gene: A Journey Through African American Culinary History in the Old South*. New York: Amistad, 2017.

Walker, Alice. *In Search of Our Mothers' Gardens: Womanist Prose*. Orlando, FL: Houghton Mifflin Harcourt, 2004.

White, M. M. *Freedom Farmers: Agricultural Resistance and the Black Freedom Movement*. Chapel Hill: University of North Caroline Press, 2018.

Williams-Forson, Psyche. "More Than Just the 'Big Piece of Chicken': The Power of Race, Class, and Food in American Consciousness." In *Food and Culture*, 121–132. New York: Routledge, 2012.

15 Disrupting the Narrative: Youth Storytelling for Food Justice

Laurel Bellante, Megan A. Carney, Deyanira Ibarra, Mary Beth Jäger, Nelda Ruiz, Claudio Rodriguez, Tommey Jodie, Kaleigh Brown, Rezwana Islam, and Amrita Khalsa

Introduction

Since being named a UNESCO City of Gastronomy in 2015—largely in honor of its four thousand–year history of continuous agricultural production as well as its rich culinary practices represented by Native, Hispanic/Latinx, and immigrant populations—the city of Tucson, in southern Arizona, has prioritized food as a key engine for economic development. While the UNESCO designation and associated efforts at "gastrodevelopment" have bolstered food-based tourism and high-end cuisine, they have also prompted debate around cultural heritage and stoked concerns about gentrification, displacement, and food injustice (Bellante et al. 2022; Kinkaid 2023; Platts 2023).

The "Future of Food and Social Justice Youth Storytelling Project," or FFSJ (2022–2023), was developed in part as a response to present debates related to food and agriculture in the southern Arizona borderlands region. The project foregrounds youth storytelling and antiracist, feminist approaches to knowledge production. Uplifting the stories of young people experiencing and witnessing food injustice, FFSJ seeks to address systems of narrative erasure and to nourish a future of food justice (Reese 2019). Storytelling is mobilized as a tool to work against the ongoing enclosure of "Tucson's Food Story" while also cultivating an "ethic of careful difference" that attends to harm and fosters a diversity of food narratives (Platts 2023). Aligned with this section's theme of encounters, we offer storytelling as a tool to foment counterhegemonic encounters and shift how we think about, narrate, and attend to food justice challenges. This chapter, cocreated with FFSJ participants, reflects on our first year with the project. Together, we explore the power of storytelling as a counterhegemonic practice that can disrupt dominant narratives around food system change and ensure that systematically excluded and oppressed communities can exercise greater agency in the pursuit of food justice. We emphasize the key role of youth, not only in making visible food injustices but also in articulating visions for more liberatory food futures.

Story-Based Strategies for Nourishing Food Justice

Stories can shift our internal and external worlds. Whether consciously or not, each day we make sense of ourselves, our lives, and our worlds through the stories we tell and the stories that circulate around us. Because communication is power, stories are not just symbolic but have real material ramifications in how decisions are made, resources are distributed, and ultimately, how lives are lived.

In the case of food system change, dominant narratives are often informed by white, middle- to upper-class voices and focused on neoliberal strategies that grow food entrepreneurship and satiate white desires to "eat the Other" (Joassart-Marcelli 2021; Hall 2020). These dynamics extend to the Tucson region, where food stories are actively enrolled for the purposes of economic development amid an omission of the voices of Indigenous, Latinx, Black, and other people of color as well as youths, low-income populations, and others fighting for food sovereignty and food justice (Platts 2023; Kinkaid 2023). Recent scholarship evaluating Tucson's UNESCO City of Gastronomy designation draws connections between the food stories that are told and broader trends in place-branding and urban development, including gentrification and displacement (Platts 2023; Kinkaid 2023). In other words, storytelling has implications for food justice. It is therefore essential that we address the imbalances, silences, and absences in who talks, who listens, and the attendant symbolic, embodied, and material impacts of these practices.

A demographic frequently absent in Tucson's food stories are the voices of young people. Literature on food justice similarly lacks engagement with youth voices. When youth do appear, they are often described as victims of food injustices such as food insecurity and obesity (notable exceptions include Leung et al. 2017; Schell et al. 2023; Trinidad 2012). Yet, from their range of experiences as children of immigrant farmworkers, food service workers, inheritors of Indigenous and other cultural wisdom, first-generation college students, community organizers, and much more, youth are often at the frontlines of our food system and offer valuable perspectives and solutions for building food justice.

Countering Archival Erasure and Seeding Alternative Futures

As critical race and anticolonial scholars have advocated, making "space for counterstories" (e.g., Dixon 2015; McKittrick 2006; Razack 1998) is a critical strategy for contesting dominant perspectives of who "speaks," who "knows," and to what ends (Kepkiewicz et al. 2015). Circulating counterstories can disrupt dominant narratives, counteract archival erasures, and seed alternative futures wherein there is greater representational justice and communication sovereignty (Dutta and Thakar 2019). As articulated by Māori scholar Linda Tuhiwai Smith (2005), "To hold alternative histories is to hold alternative knowledges. The pedagogical implication of this access to alternative knowledges is that they can form the basis of alternative ways of doing things," (p. 109). Since our project's inception, we have sought to attend to the "communicative erasures" of Black, Brown, youth, and other voices that have been absent in food stories (Pine and de Souza 2024) and uphold the autonomy and self-determination of these groups in shaping just futures.

FFSJ invites storytelling in any form and on any food-related topic. This openness allows stories to emerge organically, contributing to radical history- and future-making. This approach supports greater self-determination and representational justice. Hence, we operationalize our open approach as a kind of insurgent practice that stands in contrast to dominant forms of scholarship and official archives. Making space for this creativity and self-determination is integral to our ability to enact new imaginaries, other ways of being, and other world possibilities (Brown 2017).

Our project challenges the hegemonic whiteness of food as it exists in multiple spheres: of "alternative" food and farm spaces, of food studies as a "white space," and within the context of the neoliberal (colonial) university (Carney 2022; Garth and Reese 2020; Guthman 2011; Pine and de Souza 2024). Any effort toward food justice must not only acknowledge the expertise, knowledge, and experiences of minoritized people but also center forms of knowledge production and dissemination that allow those who have been systematically silenced to communicate in the ways that feel most appropriate and accessible (Archibald 2008; Mohanty 2003; Sandoval 2000; Trouillot 2015). At heart, these strategies are about recognizing the multiplicity of knowledges based on lived and embodied experiences; promoting the psychosocial well-being of marginalized youth, particularly queer, trans, Black, Indigenous and/or people of color (QT BIPOC); disrupting dominant framings of problems and solutions; and seeking reparations for, restoring, and upholding the dignity of people who have endured the violence of genocidal settler colonialism, transatlantic slavery, and neoliberal racial capitalism (Mohanty 2003; Pine and de Souza 2024; Tuhiwai Smith 2021; Vizenor 1994). In summary, storytelling serves as a practice both of "archival activism" and of futurity or future worldmaking that aligns with community-centered knowledge production and pathways to food justice (Flinn and Alexander 2015).

Methodology

FFSJ was launched by the Center for Regional Food Studies at the University of Arizona in the fall of 2022 to create food justice stories with, for, and by young people and their communities. The project was initiated by two white women committed to fulfilling the university's obligations as a land grant and Hispanic-serving institution, and we have sought to decenter white experiences and foreground the leadership and voices of QT and BIPOC youth and mentors. To do this, we first established a paid, six-person steering committee of community organizers and scholars, four of whom are Latinx and/or Native. The steering committee played a central role in drafting the call for applications, curating stories, leading workshops, and evaluating project activities. The initial call for applications was circulated across the University of Arizona campus, and fourteen students from distinct disciplinary backgrounds were selected. Of the fourteen interns, five identify as Native, four as POC, and eight as queer and/or trans.

FFSJ pairs interns with food justice mentors and provides biweekly mentorship circles and monthly workshops to train students in food justice and multimedia storytelling. Mentoring circles allow for reciprocal learning and help us to collectively reflect on food justice in theory and practice. Monthly workshops cover topics such as the ethics of research and storytelling, food memory, operationalizing a racialized right to food, defining food justice, and strategies for telling and disseminating food stories.

The conveners of this work are committed to antiracist and feminist approaches to food justice praxis and knowledge production. We are aware of the peril of food justice and alternative food initiatives being problematically rooted in "white desire" to enroll POCs in a "particular set of food practices" (Guthman 2008, 433). Rather than prescribe any line of questioning or final goals for this project, we have focused on providing the resources, mentorship, and community of practice and mutual care to encourage each participant to make known the stories that matter most to them, in the medium of their choosing and for dissemination in the forum that feels most relevant.

By engaging in a multistep, collaborative process to writing and analysis, this chapter is grounded in the coproduction of knowledge. First, mentors and interns wrote about our experiences in the project and circulated reflections for all to read. Next, we identified salient themes related to food justice, storytelling, and the internship program overall and used them to draft this chapter. In what follows, we share insights from our collaborative process of reflection and analysis and highlight the ways in which a community of practice in storytelling can help advance food justice.

Preliminary Reflections on the Storytelling Project

Throughout our work on this project, it became clear that the process behind creating and sharing stories is as vital as—and perhaps even more important than—the end products, the stories themselves. Here, we draw on the stories and reflective writings of our coauthors to demonstrate how storytelling practices can help advance food justice work. Within this initial pilot year of FFSJ, we observed how our storytelling process and community of practice can contribute to three key dimensions of food justice: (1) reclaiming food histories, relationships, and practices; (2) building critical consciousness and community knowledge about the food system; and (3) envisioning possibilities for more-just food futures.

Reclaiming Food Histories, Relationships, and Practices

The stories and discussions generated through FFSJ demonstrate the many ways food is not only a site of ongoing harm, exclusion, and suffering but also the place from which greater consciousness, connection, and solutions can emerge. Through storytelling, we have been able to articulate what has been ignored or silenced while also making space for narratives that communicate the power, healing, joy, and connection that is derived from food and food practices. As one coauthor observes, "Overall, all the stories have an overarching theme of community and how we have found it through food" (Tommey Jodie).

Together, we have reflected on the deep anxieties we feel about the many food challenges we face (i.e., climate change, biodiversity loss, hunger, food apartheid, chronic illness, racism, etc.). Yet, as the reflections featured here illustrate, food is also the locus for reclaiming connections, deepening one's identity, protecting traditions, defending land, water, and seeds, and seeking new ways of being with each other and the world.

Tommey Jodie (intern, twenty-one years old, Diné) is a storyteller/activist/artist/student originally from Teesto, Arizona, but raised in Winslow, Arizona. She is working toward her bachelor's degrees in food studies and nutrition and food systems.

> Growing up in a border town of the Navajo Nation, I spent countless summers with my cousins on the reservation, where we explored the land on our horses and made so many fond memories. We would travel throughout the Southwest and lower Midwest to watch and compete in rodeo. Both my late maternal and paternal grandfathers started this rodeo tradition. Them sharing their passion and traditions with me has been the greatest gift and has connected me to my culture and people.
>
> In my travels, I've seen many reservations struggling to address the growing problem of food apartheid. As I've grown older and gained more knowledge about (food) colonialism, my interest in helping overcome this epidemic has only become stronger. All the work I do is to help reclaim my people's land and traditional foodways. I am inspired and guided by so many scholars, people in my community, and my ancestors to do this work. I feel immense responsibility to reciprocate the same love and protection towards future generations that my ancestors did for me. I believe that the incorporation and legitimization of Indigenous knowledge systems is crucial and profoundly impacts both our survival and the visibility of Native communities which have been historically discarded by western ways of knowing.

Mary Beth Jäger (mentor and steering committee member, thirty-nine years old) is a queer, Xicana, Citizen Band Potawatomi, and of German descent.

> 'I want to do something involving my tribe's traditional foods.' Hearing that statement from some of the mentees I worked with filled my heart with joy and hope. This desire to connect with their foods and lands . . . is what justice looks like: young Indigenous people continuing traditions that people have wanted, and still want to, stomp out.

Building Critical Consciousness About the Food System

By creating spaces to share stories with people from a range of life experiences and trajectories, we can build consciousness about the food system and how it relates to power, privilege, and violent legacies of colonialism, racism, and oppression. Through a predominantly nonwhite community of praxis, participants were able to explore how whiteness pervades the food system and reflect on our positionality. The stories we have shared reveal the innumerable ways that food justice can be conceptualized and reflected on through story. As one coauthor observes, "There's no one story of food, justice, or food and justice. We all have a different experience of it" (Amrita Khalsa).

Rezwana Islam (intern, twenty-four years old) is pursuing a B.S. in public health and a B.A. in creative writing. Raised in Tucson, she is one of three children of Bangladeshi immigrant parents and draws upon her experience being queer, Muslim, and neurodivergent.

> I grew up visiting Indian, Middle Eastern and Asian markets and eating Bengali food cooked by my parents for dinner, American food like pizza or sandwiches for lunch while being surrounded by foods unique to southern Arizona. Not always eating the same thing as my friends or having favorite foods they had never heard of brought a feeling of isolation. Despite this, exploring the shelves of tiny local markets and picking snacks or desserts from Asia and the Middle East to eat in the car on the way home brought a sense of joy.
>
> A natural love for the environment and public health combined with a love for community and social justice manifested in a desire to know more about the connections people make with the earth. Becoming aware of how the health of the earth affects different foods produced by different regions and cultures sparked my interest in studying food cultures and using storytelling to explore my own relationship with food and the community around me.

Kaleigh Brown (intern, twenty-one years old) is a recent graduate in film and television production and environmental studies. She was raised in a white, middle-class family in Tucson.

> I grew up amongst the unmistakable sights, smells, and tastes of a health food store, specifically the New Life Health Food Center on Speedway Boulevard in Tucson, Arizona. It was routine for my grandpa to pick me up from school in his dusty Toyota pickup truck and drive us to New Life where we would wait impatiently for 8 p.m. when my step-grandma, the store manager, could lock up for the night. If boredom struck, I would embark on grocery adventures. A table by the entrance piled high with books on herbal medicines and vegan diets provided pictures. The poor lighting at the back provided the perfect opportunity to sneak around aisles like a spy while deciphering names of teas stocked against the wall. Finally, 8 p.m. would arrive. My step-grandma would pull out her keys and lock the doors, jiggling each handle. With my head on my hands and my elbows resting on the counter, I would watch as she worked the register. Click. Click. The numbers would eventually match up, meaning we were excitedly closer to going home.

> Now, I take Speedway Boulevard on my daily commute to college. Where New Life should be, there is now nothing but empty windows. The family running the store closed its doors in 2017. National grocery chains were popping up everywhere, many with health and wellness selections, becoming fierce competitors to small, family-owned health food stores. On one hand, spaces like New Life have largely disappeared and are something most people can no longer experience. On the other hand, for many, access to such stores has always been out of reach. Such stores were not only few and far between but were often more expensive than chain stores. My heart aches to once more dance down the aisles with my step-grandma after locking up for the night. But, I long even more for everyone to have more equitable access to the foods once found in stores like New Life.

Amrita Khalsa (intern, twenty-nine years old) is a recent graduate in food studies. She is of Swedish and Texan heritage and was raised in Tucson in an upper-middle-class neighborhood.

> I grew up on the northeast side of Tucson with access to fresh produce. My parents fed my three siblings and me a vegetarian diet, guided by an ayurvedic yoga lifestyle and the alternative, organic, and slow food movements. This informed what food I consume and how I connect food and wellness. I always knew this was quite different from the western "American" diet, but what I didn't know was that the fresh foods my family ate were not as accessible in other neighborhoods.
>
> I have come to realize how government policy has made cheap, processed food so prominent and accessible, as opposed to fresh produce. It wasn't until my food studies program that I began to understand food justice topics including the importance of equitable food access and cultural foodways. Before, I didn't realize the extent to which food access was and still is political and why I see nicer grocery stores in areas I grew up in, and more fast-food chains and liquor stores in other areas.

Our commitment to ethical storytelling and openness to a range of storytelling formats also support a community-centered approach to knowledge production. Reflecting on the importance of telling stories in different formats, Mary Beth Jäger (mentor/coauthor) writes,

> I enjoy listening to students describe how they want to tell the story of their tribe's traditional foods in creative ways, whether it be through oral storytelling or art. I think these mediums make stories more accessible to share with other community members than traditional academic methods. I'm grateful this internship gives students the space to be creative, to tell their story in their own way and not be hemmed in by the ivory tower.

Nelda Ruiz (mentor and steering committee member, thirty-six) is a *fronteriza* community organizer, cultural worker, and educator born in Ambos

Nogales in the Sonoran Borderlands. She has lived in Tucson since 2001. Reflecting on the power of controlling our own narratives, Nelda writes,

> Sharing stories and sharing space creates impactful moments of cohesion and trust. In a world where "outsiders" too often extract stories from our communities and retell them as their own or use them in sensational ways, it is critical that we tell our own stories in our own ways. Food stories are one way we can do that. Food becomes a "vehicle" for uncovering our different experiences, struggles, and memories of joy. I've been working alongside and learning from students through a mentorship relationship, having conversations about the meaning of food justice, the importance of ethical documentation, and why and for whom a story should be told. Throughout these conversations, my underlying message is always to inspire students (and everyone) to control their narrative and share their own experiences of the food system. What was crucial was to create an understanding that a story is a tool, and that tool can serve a function to the community you belong to.

Envisioning More-Just Food Futures

By capturing both quotidian moments and expressions of broader systems of violence, storytelling can link lived experiences to the systemic changes that are needed. By making known the embodied and affective experiences of food injustice, stories can help cultivate collective action toward more-just food futures.

Claudio Rodriguez (mentor and steering committee member, thirty-seven) is a community organizer, artist, and storyteller. He migrated with his working-class family from Mexico and was raised in the barrios of the south side of Tucson. In describing how storytelling changes the narrative, Claudio writes,

> I have been part of the food justice movement for over 15 years. Storytelling has been one of my most useful tools to develop understanding, empathy, and inspiration to change the food system we live in. Our stories don't have to be tied to specific struggles or grand systemic changes; our stories can be those little moments of how a double cheeseburger brought us joy and a sense of success when fitting into the American dream. Our stories can also be the change of seasons that our lives have taken, for example my growth from gangs to gardens, how my trauma was healed by planting seeds to grow corn, how my hands in the dirt became my purpose.
>
> Being part of this mentorship has blessed me with challenges and opportunities in guiding my mentees on best practices of storytelling, why it matters and how we show up in this work. It always helps to have someone you can work through ideas of how a story can and should be told. A common concern that came from my mentees was, "should I share stories that seem too controversial?"

My recommendation was always yes, because as storytellers it is our responsibility to share perspectives that are often left out of conversations, to share ideas, hopes and dreams of what our relationship to food can look like. My reminder to youth in storytelling is that stories will help you understand yourself and give others the courage to share their truth. These are nurturing practices that help heal our environment and ourselves. This is why we share these stories.

Deyanira Ibarra (research assistant, Ph.D. student in sociocultural anthropology) is a queer Chicana, born and raised in the Santa Ynez Valley, an agricultural region of California's central coast.

Growing up in a Mexican family—lessons, tales, and family histories were circulated through stories. As most of my elders never learned to read or write, oral storytelling felt like an avenue for profound interpersonal connection. When I shared with my grandfather that I would be moving to Arizona for graduate school, this prompted him to share about his time picking lettuce in Yuma, AZ as a Bracero and in turn my own reflection on the different experiences we would have in the same state.

Participating in FFSJ both affirmed and expanded my understanding of storytelling by positioning it as a craft and a theory of change. This internship centers the experiences of BIPOC who are impacted by food injustice and foregrounds storytelling as a mode of survival and resistance. Stories are important because they help us make sense of the world, build and transform identities, and are a tool for shaping and reclaiming narratives about certain experiences or groups of people. In supporting the development of food stories, FFSJ both amplifies the expertise of young people and makes an intervention in who gets to represent food injustice. To build awareness around food injustice, storytelling is particularly salient because it can catalyze pathways to much needed cultural, political, and social change on many scales.

As demonstrated in these story excerpts and reflections, the practice of gathering and sharing stories creates ripple effects for everyone involved. The pilot year of FFSJ has transformed each of us. It has helped us to work through stories of pain and injustice, unearth examples of hope and resilience, raise our awareness of complex issues, and cultivate a shared compassion and solidarity that opens our hearts and minds to conceive of new pathways forward.

Conclusion

Tucson-based community organizer (and coauthor) Claudio Rodriguez often says, "If you're not at the table, you're on the menu." Indeed, a critical first step in ensuring a more inclusive "table" for determining the shape of food-based development projects and justice work is to center the voices and experiences of those who are at the front lines of food injustice and have been left out of the discussion, particularly youths.

Scholars have rightly called for a food justice praxis that works to address interlocking structures of oppression at the foundation of food system inequities (i.e., capitalism, patriarchy, white supremacy, colonialism) (Kepkiewicz et al. 2015). While we have argued in favor of storytelling as a tool for food justice, we are under no illusions that increasing representational justice alone is sufficient to ameliorate the challenges before us. That said, the practice of gathering, telling, and disseminating stories can counteract the violence of erasure, disrupt dominant food stories, and raise serious questions about how food stories are enrolled in neoliberal approaches to development.

The steering committee behind FFSJ has seen the role storytelling can play in fostering positive change. Sharing stories and experiences of food injustice (i.e., lack of green infrastructure, the disconnection from our traditional foods, lack of access to healthy, affordable foods, etc.) has been a vital part of over twenty years of grassroots work in neglected areas of Tucson that has culminated not only in the expansion of green spaces such as backyard gardens, shaded walkways, and community gardens but also increased access to the resources required to fund and sustain these spaces. These are but a few examples of how stories can influence material change. In other words, seeds, dreams, and stories all play a role in creating system-level change that can help address the root causes of food insecurity and injustice.

In sum, while storytelling in itself is insufficient to address the range of challenges we face, it should be considered a vital tool for working toward food justice. Storytelling is foundational for shifting consciousness, challenging Anglo-normative assumptions, and counteracting erasures of not only the food-related traumas experienced by POC and other historically marginalized/resilient groups but also our food-based traditions, joys, and pleasures. Keeping this in mind, we plan to continue building on the initial work produced in the pilot year of FFSJ to encourage more storytelling practices and stories to be shared in our region, and we hope to inspire others to undertake their own storytelling initiatives. The stories created in this initial year of FFSJ are now available for circulation as an edited compilation (FMI, visit https://crfs.arizona.edu/). Through centering and empowering the voices of youth, we ensure that the up-and-coming generations have the skills, relationships, and confidence to share their stories widely and take their place at the table of dreaming and enacting more-just food futures.

References

Archibald, J. A. 2008. *Indigenous Storywork: Educating the Heart, Body, Mind and Spirit.* Vancouver, Canada: University of British Columbia Press.

Bellante, L., Carney, M. A., and Owen, G. 2022. "Leveraging university resources to build awareness, support regional food policy, and disrupt dominant narratives guiding food-based development: Examples from University of Arizona's Center for Regional Food Studies." *Journal of Agriculture, Food Systems, and Community Development* 11(3): 9–11.

Brown, A. M. 2017. *Emergent Strategy: Shaping Change, Changing Worlds.* Chico, CA: AK Press.

Carney, M. A. 2022. "Whiteness and settler colonial logics in the Pacific Northwest hops and craft beer industries." *Food, Culture & Society* 26(5): 1032–1055. https://doi.org/10.1080/15528014.2022.2069440.

Dixon, B. A. 2015. "Rewriting the call to charity: From food shelf volunteer to food justice advocate." *Journal of Agriculture, Food Systems, and Community Development* 5(2): 71–79.

Dutta, M., and Thakar, J. 2019. "'Communication sovereignty' as resistance: Strategies adopted by women farmers and the agrarian crisis in India." *Journal of Applied Communication Research* 47(1): 24–26.

Flinn, A., and Alexander, B. 2015. "Humanizing an inevitable political craft: Introduction to the special issue on archiving activism and activist archiving." *Archival Science* 15: 329–335.

Garth, H., and Reese, A. M. (eds.). 2020. *Black Food Matters: Racial Justice in the Wake of Food Justice.* Minneapolis: University of Minnesota Press.

Guthman, J. 2008. "Bringing good food to others: Investigating the subjects of alternative food practice." *Cultural Geographies* 15(4): 431–447. https://doi.org/10.1177/1474474008094315.

Guthman, J. 2011. "If they only knew: The unbearable whiteness of alternative food." In A. H. Alkon and J. Agyeman (eds.), *Cultivating Food Justice: Race, Class and Sustainability*, 263–281. Cambridge, MA: MIT Press.

Hall, B. 2020. "Soul food gentrification: Food, racial heritage tourism, and the redevelopment of Black space in Miami." In H. Garth and A. M. Reese, *Black Food Matters:*

Racial Justice in the Wake of Food Justice, 158–180. Minneapolis: University of Minnesota Press.

Joassart-Marcelli, P., 2021. *The $16 taco: Contested geographies of food, ethnicity, and gentrification.* University of Washington Press.

Kepkiewicz, L., Chrobok, M., Whetung, M., Cahuas, M., Gill, J., Walker, S., and Wakefield, S. 2015. "Beyond inclusion: Toward an anti-colonial food justice praxis." *Journal of Agriculture, Food Systems, and Community Development* 5(4): 99–104. https://doi.org/10.5304/jafscd.2015.054.014.

Kinkaid, Eden. 2023. *Consuming the Creative City: An Urban Political Ecology of Gastrodevelopment in Tucson, Arizona.* Doctoral dissertation. University of Arizona.

Leung, M. M., Agaronov, A., Entwistle, T., Harry, L., Sharkey-Buckley, J., and Freudenberg, N. 2017. "Voices through cameras: Using photovoice to explore food justice issues with minority youth in East Harlem, New York." *Health Promotion Practice* 18(2): 211–220.

Mohanty, C. T. 2003. *Feminism Without Borders: Decolonizing Theory, Practicing Solidarity.* Durham, NC: Duke University Press.

Pine, A., and R. de Souza. 2024. "Hunger, survivance, and reparative food policy: A racial analysis of the 'right to food.'" In S. E. Dempsey (Ed.), *Organizing Eating: Communicating for Equity Across U.S. Food Systems.* New York: Routledge.

Platts, Ellen. 2023. *Future-Making in the City of Gastronomy: Food Heritage and the Narrative Commons.* Doctoral dissertation. University of Maryland.

Razack, S. 1998. *Looking White People in the Eye: Gender, Race, and Culture in Courtrooms and Classrooms.* Toronto: University of Toronto Press.

Reese, A. M. 2019. *Black Food Geographies: Race, Self-Reliance, and Food Access in Washington, DC.* Chapel Hill, University of North Caroline Press.

Sandoval, C. 2000. *Methodology of the Oppressed.* Minneapolis: University of Minnesota Press.

Schell, E. E., Winslow, D., and Shrestha, P. eds. 2023. *Food Justice Activism and Pedagogies: Literacies and Rhetorics for Transforming Food Systems in Local and Transnational Contexts.* Lanham, MD: Lexington Books.

Trinidad, A. M. 2012. "Critical indigenous pedagogy of place: A framework to indigenize a youth food justice movement." *Journal of Indigenous Social Development* 1: 1–17.

Trouillot, M. 2015. *Silencing the Past: Power and the Production of History.* Boston: Beacon Press.

Tuhiwai Smith, L. 2005. Imperialism, history, writing, and theory. In *Postcolonialisms: An Anthology of Cultural Theory and Criticism*, 94–115.

Tuhiwai Smith, L. 2021. *Decolonizing Methodologies: Research and Indigenous Peoples*. London: Bloomsbury Academic.

Vizenor, G. 1994. *Manifest Manners: Post-Indian Warriors of Survivance*. Hanover, CT: Wesleyan University Press.

IV Intersections

16 New Openings for Worker Justice in the Food System: A Call for Centering Labor in Food Justice

Teresa M. Mares and Laura-Anne Minkoff-Zern

Introduction

At periodic moments in US history, the plight of food workers and farmworkers has peaked national attention. From the publication of *The Jungle* by Upton Sinclair, to the broadcast of the *Harvest of Shame* documentary, through the United Farmworkers' grape boycotts, and most recently, during the COVID-19 pandemic, we have been repeatedly reminded of the harsh realities of food and farm work. Despite these moments of visibility, we still depend on a food system that perpetuates violence and injustice for workers in both public and private spaces.

As researchers who are inspired by national coalitions like the Food Chain Workers Alliance and who have studied food justice for a combined four decades, we support the notion that the food justice movement must center labor to nurture a just and sustainable food system (Food Chain Workers Alliance and Solidarity Research Cooperative 2012; Liu and Apollon 2011). As we argue in our coauthored book, one cannot solve the problem of unjust labor by focusing on a singular labor process or sector in the food system (L.-A. Minkoff-Zern and Mares 2025). The same tomatoes that are picked by undocumented immigrants, are sliced by underpaid workers in fast-food kitchens, rung up by Walmart cashiers denied the benefits of collective bargaining, and fed to hungry children by women facing the pressures of holding a paid job while simultaneously being held responsible for their family's consumption.

We must also recognize that the US food system is founded upon slavery's legacy; without it and the brutal labor systems that followed, plantation agriculture and the normalization of worker exploitation and industrial agriculture would not exist. As Sbicca and colleagues describe in this section, the violence of plantation economies has persisted in carceral food systems today. And yet, whether it is the unpaid intern at the small diversified farm or the migrant workers sustaining large-scale monocrop agriculture, food injustice is as possibly present in biodynamic kale at the farmers market as it is in milk coming from corporate megadairies. Labor injustice is not merely a question of scale or size, as alternative or localized food systems are not free from exploitation and violence (Gray 2013).

As agriculture and food companies have consolidated both vertically and horizontally over the past half a century, new strategies have developed to keep wages low and workers in replaceable positions. Across the food system, we have seen a move away from full-time work with stable schedules and benefits to part-time and erratic employment, as well as outsourced and subcontracted labor. In specific sectors, such as service, the franchise model has distanced the corporate leadership from the everyday laborers handling our food. In food delivery, we have seen an increase in the gig economy, where workers are categorized as independent contractors and denied the rights of minimum wage and days off. And, across the food system, automation is gaining increasing traction, though we remain skeptical that robots will simply or quickly replace the human labor required to make and serve food.

Alongside these transformations, we have seen a growth in food justice movements and scholarship over the past several decades. Many food justice organizations focus on developing small-scale alternatives or centering BIPOC leadership of food-related initiatives, and while labor has been a focus of some food justice activism and research, it has not been at the center of the movement (see Agyeman and Alkon 2011; Gottlieb and Joshi 2010; Sbicca 2018). Given this incomplete framing, researchers have thus far missed the opportunity to create a powerful avenue for creating structural social and environmental change and helping to nurture food justice. We believe that the food justice movement would be strengthened through more fully embracing workers' rights as a top priority. In this chapter we look at both the structural exploitation of food workers, as well as their resilience and potential for systemic change along the food chain to reveal new openings for the broader movement for a just food system.

What Do We Know About Food and Labor?

There is a long history of racialized workers across food and agriculture sectors being denied basic rights and safety protections while simultaneously

Table 16.1
2023 BLS food system data

Sector	No. of people employed	Hourly mean wage ($)	Annual mean wage ($)
Production	711,660	19.34	40,225.91
Processing	879,490	18.59	38,666.67
Distribution	3,608,930	20.32	42,271.33
Retail	1,673,620	15.66	32,573.33
Service	9,655,990	16.24	33,772.50
Waste	473,479	22.05	45,862.11

being exposed to toxic and strenuous working conditions. Essential food worker positions have a high representation of undocumented people of color in comparison to the general population (chapter 17, this volume; Pangborn and Rea 2020; Ramos et al. 2020). People of color make up 50 percent of paid essential workers in food and agricultural sectors. This overlaps with class dynamics, as 70 percent of essential workers overall also do not have a college degree. Workers in these sectors are also less likely to be in a union and make a lower median wage than nonessential workers (McNicholas and Poydock 2020).

The data in table 16.1 from the Bureau of Labor Statistics (BLS) show us the size and scope of each sector in the food system as of 2023. These data capture frontline workers in the food system, not including managers, supervisors, and white-collar workers. We note that these data do not include undocumented workers, independent contractors, or gig workers and that these data are not available for food-related labor in the home given that it is unpaid.

Exploitation of food sector workers is common and occurs in a variety of ways. In response, labor scholars have studied the impacts of economic fissuring, worker disenfranchisement, and organizing strategies (Jayaraman 2013; Levine and Striffler 2015; Rosenblum 2017). Feminist scholars have highlighted the unequal and undervalued reproductive labor burden that women shoulder (Adams 1995; Sachs and Allen 2007; Bowen, Brenton, and Eliott 2019; Carney 2015; Cairns, Johnston, and MacKendrick 2013; Mares 2017). Yet even in the interdisciplinary fields of food studies and food systems, scholars tend to look at food labor in separate food sectors without making linkages between them, such as farm work, service work, home food preparation, school food work, and so on (Besky 2013; Gaddis 2019; Gray

2014; Kurtz et al. 2019; Shreck, Getz, and Feenstra 2006). In what follows we provide examples of compelling worker organizing efforts that help us to connect shared concerns and opportunities across food sectors.

A Food Justice Movement Led by Workers

Despite the entrenched injustice in our food system, we are inspired by the contributions and ongoing struggles of food system workers. As we write this, we are witnessing a revival of worker organizing that hasn't been seen for decades. Newer forms of worker-led organizing such as those pursued by members of the Worker-Driven Social Responsibility Network offer new approaches to worker justice, as discussed by Vera Chang in chapter 17. This renewed energy among workers to collectively organize is a response to the COVID-19 pandemic, paired with increasingly low wages in relation to rising costs of living (McCallum 2022). Of course, this growth is happening unevenly across the sectors of food system work, and different kinds of organizing methods are more or less useful for certain groups of workers. But it is happening, and that is evidence that there is reason to be hopeful. In what follows, we provide organizing examples from three food sectors: union organizing in food retail, state-level legislation in food production, and The Fight for $15 movement in food service.

Organizing Unions in the Grocery Industry

Despite megastore mergers and consolidation in the grocery sector and increasing precarity among retail workers, there is a long history of unionization for today's food workers to draw upon. What had once been a highly unionized sector is now more reflective of the rest of American retail, with low levels of unionization and increasingly precarious employment. Yet the past few years of unionization in retail and other food sectors show that the tide may be changing. Currently, public perception in favor of unions is at the highest level since 1965, and hopefully, this will translate to reversing, or at least slowing, the downward trend in unionization (McCarthy 2022).

The original mom-and-pop grocery stores that preceded large-scale chains were allied with unions, as they were typically run by the family alone. In the early 1900s, when there was a push for longer operating hours in the grocery industry, family-run stores fought back along with unions, as they wanted to maintain their nonwork time. Specialty food trade unions, such as the Amalgamated Meat Cutters Union and bakery unions, and other retail unions also saw their professions being swallowed by the grocery industry as it beat out small specialty food shops and supported the grocery union struggles. As expected, large chain operations resisted unionization, both with carrots (higher wages and pension plans) and sticks (threatening to fire employees and closing unionized stores and warehouses). As there were no legal protections for workers to unionize until the National Labor Relations Act was passed in 1935, only a small number of grocery chains were unionized before this time (Levinson 2019).

While large chains such as A&P initially fought unionization in their stores, the American Federation of Labor (which later became the AFL-CIO) and others eventually won this struggle after an explosive battle, and by the 1960s, grocery retail became one of the most highly unionized sectors, with about 40 percent of all retail food workers being organized (compared to 12 percent of all retail workers). At that time, all of the large chains were highly unionized, mostly by the Retail Clerks International Association and three other major unions—the Meat Cutters, the Teamsters, and the Retail, Wholesale, and Department Store Union. These unions aligned by refusing to cross picket lines, as they represented food workers in other sectors as well, such as trucking and warehouse work. This kind of cross-sector support was crucial to successful food retail organizing (Milkman 2022b).

Relative to other private industries, the grocery sector still has a relatively high proportion of unionized labor. In 2019, only 6 percent of the private sector was unionized, with only 4 percent of retail workers in unions. Comparatively, 16 percent of grocery workers were unionized (Milkman 2022a, 2022b). While this is still low compared to unions' height in the 1960s, it is significant, and is potentially growing as part of the postpandemic labor movement (DiNatale and West 2022). Grocery unions have survived relative to other retail unions due to the place-bound nature of the industry. Further, the corporate deregulation of the 1970s, which significantly affected other sectors of the food chain, such as trucking, had less of an impact on the grocery sector (Milkman 2022b).

Historically, high rates of unionization have meant better jobs for food retail workers. Stable full-time jobs, seniority rights and promotion systems, procedures for grievances, protection against arbitrary firings, and higher wages all resulted from union organizing. Part-time work and erratic scheduling were regulated and restricted. Unlike today's largely nonunion retail workforce, organized grocery workers had good pension plans and affordable health and dental coverage.

In the 1980s, rates of unionization were generally taking a downturn, grocery workers included. The expansion of Walmart as a global retailer of food played no small part in that decline for the grocery sector, as their antiunion tactics drove down labor standards for the industry. Unfortunately, many other grocery retailers have followed Walmart's lead. In 2003, Safeway workers staged the longest grocery strike in US history. For nineteen weeks, workers stayed off the job fighting cuts to wages and benefits. Safeway's so-called competitors, Albertsons and Krogers, agreed to also lock out their union workers in solidarity with Safeway, and the companies shared their losses and revenues to weather the strike. Ultimately, the grocery chain(s) won, arguably signifying a shift in the industry from middle-class employment to a low-wage job (Howard 2021; Jayaraman 2014).

Fortunately, however, unionization and labor organizing among retail and other workers is on the rise. The COVID era, which has increased the visibility of frontline workers, and the related economic downturn, has led to increasing political support of unions (DiNatale and West 2022). While the growth in unions has been slow, a more grassroots shift in worker identity and social movement organizing has been growing as workers have been deemed essential workers (McCallum 2022). What this means in the longer term for grocery workers is still unknown, but organizers are thus far succeeding in taking advantage of this moment, with major victories at Amazon, Starbucks, and others.

As with other sectors of the food system, food retail employment does not inherently have to equal bad or undesirable jobs. Food groceries started out in the US as community- and neighborhood-based stores, which offered families stable income and respectable roles in their diverse communities. Many forms of resistance as well as alternatives have been persistent in attempts to reform food retail as we know it—from fighting back against superstores like Walmart to creating worker-owned cooperative groceries. We must keep looking to these models as we also look for solutions to food injustice and connect the dots between food access and exploitative retail labor.

State-Level Campaigns for Farmworker Rights

Unlike grocery workers, farmworkers do not have the federally protected right to form a union and have had to find more state-based strategies and other approaches to fight for better wages and conditions. Farm laborers (and domestic workers) were explicitly excluded from the National Labor Relations

Act of 1935, which protects workers' right to form a union, meaning they can be legally fired for organizing one in most states to this day.[1] This is where state-level organizing, consumer-focused campaigns, and other kinds of worker organizing have a significant role to play, such as the Coalition of Immokalee Workers and the Worker-Driven Social Responsibility model, discussed in chapter 18. The exclusion of farmworkers from traditional labor organizing and union membership has led to a long-standing absence of agricultural interests in food system union organizing. Additionally, given the racialized history of farm labor, especially in the South and the West, some labor unions were averse to including workers based on racial difference. Although there were early efforts to coordinate unionizing efforts across food sectors, such as field work and canneries (processing), ultimately, farmworkers have been left behind in formal unionization efforts on a national scale (chapter 17, this volume; Walker 2004).

Despite the historic and structural inequalities underpinning the struggle for just treatment of farmworkers on a federal level, there have been many successes at the state level. Perhaps the most well-known are the accomplishments in California, instigated by the organizing work of the United Farm Workers Organizing Committee (which eventually became the UFW). In 1965, Filipino grape workers and members of the AFL-CIO-affiliated Agricultural Workers Organizing Committee in Delano, California, started striking against low wages and dreadful working and living conditions. They soon asked a newly formed, predominantly Latinx union, the National Farm Workers Association, to join them, and together they became the United Farm Workers. To gain national attention and support the strike, Cesar Chavez, Dolores Huerta, and other organizers orchestrated a march on Sacramento, California's capital. Over the following four years, the UFW and its collaborators organized the largest food boycott in US history. By 1970, all table grape workers were under union contract, and by 1977 the UFW had secured over a hundred union contracts. These contracts guaranteed better wages, benefits, and working conditions for farm labor (Ganz 2009; Minkoff-Zern 2017). In response to the national-level grape boycotts and worker strikes organized by workers and advocates, California passed the California Agricultural Labor Relations Act of 1975, the first state-level labor law establishing the right to collective bargaining (to form a union) for farmworkers. This state law created rules and regulations similar to the National Labor Relations Act. In the following years, the UFW was crucial to getting specific legislation passed to protect farmworker rights at the state level, playing a large role in helping pass the Agricultural Labor Relations Act and establishing California's Agricultural Relations Board in 1975 (Martin 2003).

Since the passage of California Agricultural Labor Relations Act of 1975, several other states have passed legislation that directly addresses the systematic exclusion of farmworkers from federal labor protections. Many laws have been passed in the last few years, perhaps evidence of a new wave of attention to food justice broadly and labor injustices in the food system specifically over the past several decades. In 2019, New York passed the Farm Laborers Fair Labor Practices Act, which now awards farmworkers overtime pay when they labor over fifty-six hours a week (to be incrementally reduced to forty hours a week by 2032, when it will match the threshold of most other US workers). This bill also included the right to collectively bargain in New York, which was also won in a state supreme court case just before the passage of the bill. Following this momentum, in 2021, two more states signed farmworker labor bills into law. Washington state passed SB 5172, a law that creates overtime protections for agricultural workers specifically. In Colorado, a more comprehensive bill, SB21-087, removed long-standing regulations exempting farm laborers from state minimum wage laws, the right to unionize, and overtime pay. The law also bans the use of a short-handled hoe, which is known to cause debilitating back problems, following other states that have banned this tool, such as California, Arizona, New Mexico, and Texas. In 2022, California passed additional farmworker legislation, making it easier for farmworkers to vote in a union election, following decades-long declines in farmworker unionizing since they gained the right in California in 1975 (Nieves 2022).

In all these state-level cases, farmworkers, advocates, and community organizers spent years mobilizing, protesting, and rallying to bring attention to the unequal system of farmworker labor laws and the historic lack of worker protections in the agricultural industry. Local and national community-based groups such as the Workers Center of Central New York, Community to Community Development (C2C) in Washington State, and the Colorado Farmworkers' Rights Coalition, among others, worked tirelessly to bring farmworkers' rights to the attention of both the broader population and policymakers. Farmworker organizing can be extremely challenging given the physical isolation where farmworkers often live and work.

Further, farmworkers, who tend to be immigrants and non–English speakers, are socially isolated even from their local communities in the US, making it harder for them to communicate, advocate, and even know their legal rights. Finally, given that such a large proportion of the agricultural workforce is undocumented, there is fear among this population of speaking up or organizing when laws are violated or practices do not meet basic regulations.

These state laws, while significant, have been passed in a piecemeal manner, with variations in agricultural labor regulations by state. What is notable about these laws is that they merely bring farmworkers up to the same basic level of rights and protections of nearly all other workers in the US, and while farmworkers in some states are now protected to organize unions, in most states they are not. While these are wins for workers, it means that thousands of farmworkers across the country still do not have these basic rights. Despite state-level progress, federal legislation is still needed for comprehensive national protections of agricultural workers.

The Fight for $15 Campaign

The Fight for $15 campaign calls for the federal minimum wage to be raised to $15 per hour, a significant increase from the current rate of $7.25, which was set in 2009 and still remains as of 2025. This movement began in earnest in November 2012 when hundreds of fast-food workers walked off the job in New York City. Although the Fight for $15 is not limited to food service workers, the movement has gained the most traction among fast-food workers, who are particularly impacted by the absence of a livable wage and barriers to worker organizing (Orleck 2018). While the Fight for $15 is currently the most visible and active movement seeking to improve the economic realities of food service workers, it has a longer history and draws upon decades of organizing efforts among low-wage workers.

Over 2014 and 2015, the Fight for $15 organized strikes and sit-ins in hundreds of US cities, bringing together thousands of fast-food workers with those in other industries. Internationally, protests over wages at fast food restaurants were also organized in the United Kingdom, Japan, India, and Brazil. The US protests during this period were often held in conjunction with, or organized in solidarity with, Black Lives Matter protests. The largest strike during this time happened on April 15, 2015, when thousands of workers walked off the job in more than two hundred US cities.

One of the biggest demands of the Fight for $15 is connected to just-in-time and erratic scheduling practices for restaurant workers, particularly those in fast food. As Orleck (2018) explains, when schedules are planned by a computer algorithm, workers are unable to plan for childcare or have any security that their bills will be paid. An important success of the Fight for $15 occurred in 2017, when the New York City Council passed legislation requiring that fast-food workers have their work schedules fourteen days in advance and banning back-to-back shifts.

Critics of the Fight for $15 movement argue that menu prices must increase in order to account for increased labor costs or that workers will receive fewer hours at the increased rate of pay. When looking at chain restaurants in particular, these arguments are merely distractions from the immense corporate profits these chains earn. For example, in 2022 alone, the net profit for McDonalds was $6.17 billion, and for Starbucks it was $3.28 billion. Some believe that the turn to automated order kiosks, like those found at McDonalds and other fast-food restaurants, has accelerated because of calls for higher pay. Indeed, automation has certainly been a strategy for cost-cutting within fast food for decades, and a fast-food hamburger made by human hands might soon be a thing of the past as robots like Flippy 2 become more commonplace.

In a 2021 report on the impacts of the Fight for $15, the Economic Policy Institute presented findings that the movement has increased the wages of twenty-six million workers, who have earned an additional $150 billion dollars of income. Of these workers, an estimated twelve million are workers of color and eighteen million are women (Lathrop, Lester, and Wilson 2021). The report unfortunately does not provide a breakdown of how many of these workers are in food service. Although the movement has not yet realized a national increase to the minimum wage, it has motivated significant gains at municipal and state levels. Numerous states, including California, Massachusetts, Maryland, New Jersey, Illinois, Connecticut, and Florida, have passed legislation that requires a gradual increase to $15 per hour, and the cities of Seattle, San Francisco, and New York City have done so as well.

Conclusion

In this chapter we have argued that the food justice movement would be stronger with a more explicit focus on food workers' rights. Our focus on unions, state-level legislation, and wage-related campaigns highlights just a few examples of ways to address these systemic inequities. Other approaches, such as making change through grassroots organizations like ROC-United, the Food Chain Workers Alliance, and the Worker-Driven Social Responsibility Network or engaging policy mechanisms like the federal farm bill and local food policy councils, also are promising ways to change workers' lives across food system. As Maggie Dickinson writes in chapter 19, on the Green New Deal, comprehensive legislation that seeks to fundamentally shift systems of inequality and injustice can be linked with movements for ecological change. These efforts are shifting norms around racially and socially just labor practices and should inspire all workers and activists that there is not only more to do to solve food system injustices but an existing framework to dive in and do it. As the introduction to this volume asserts, labor justice is an integral part of nurturing food justice and building bridges with other social movements. The time is perfect for the food justice movement to address labor head on and for the labor movement to take food justice seriously.

Note

1. In the following states, farmworkers are now protected to unionize because of state-level policies: Arizona, California, Colorado, Hawaii, Kansas, Kentucky, Louisiana, Massachusetts, Nebraska, New Jersey, New York, Oregon, Washington, and Wisconsin.

References

Adams, Jane. 1995. "Individualism, Efficiency, and Domesticity: Ideological Aspects of the Exploitation of Farm Families and Farm Women." *Agriculture and Human Values* 12 (4): 2–17.

Agyeman, Julian, and Alison Alkon, eds. 2011. *Cultivating Food Justice: Race, Class, and Sustainability*. MIT Press.

Besky, Sarah. 2013. *The Darjeeling Distinction: Labor and Justice on Fair-Trade Tea Plantations in India*. University of California Press.

Bowen, Sarah, Joslyn Brenton, and Sinikka Eliott. 2019. *Pressure Cooker: Why Home Cooking Won't Solve Our Problems and What We Can Do about It.* Oxford University Press.

Cairns, Kate, Joseé Johnston, and Norah MacKendrick. 2013. "Feeding the 'Organic Child': Mothering Through Ethical Consumption." *Journal of Consumer Culture,* 13 (2): 97–118. https://doi.org/10.1177/1469540513480162.

Carney, Megan A. 2015. *The Unending Hunger: Tracing Women and Food Insecurity Across Borders*. University of California Press.

DiNatale, Natale, and Kayla West. 2022. "Union Representation Petitions Are Up 57 Percent, but That's Not All!" *National Law Review*. May 16. https://www.natlawreview.com/article/union-representation-petitions-are-57-percent-s-not-all.

Food Chain Workers Alliance and Solidarity Research Cooperative. 2012. "The Hands That Feed Us: Challenges and Opportunities Along the Food Chain." Food Chain Workers Alliance, June 2012. https://foodchainworkers.org/wp-content/uploads/2012/06/Hands-That-Feed-Us-Report.pdf.

Gaddis, Jennifer E. 2019. *The Labor of Lunch: Why We Need Real Food and Real Jobs in American Public Schools*. University of California Press.

Ganz, Marshall. 2009. *Why David Sometimes Wins: Leadership, Organization, and Strategy in the California Farm Worker Movement*. Oxford University Press.

Gottlieb, Robert, and Anupama Joshi. 2010. *Food Justice*. MIT Press.

Gray, Margaret. 2013. *Labor and the Locavore: The Making of a Comprehensive Food Ethic*. University of California Press.

Howard, Phillip. 2021. *Power and Concentration in the Food Industry: Who Controls What We Eat?* Bloomsbury.

Jayaraman, Saru. 2013. *Behind the Kitchen Door*. Cornell University Press.

Jayaraman, Saru. 2014. "Shelved: How Wages and Working Conditions for California's Food Retail Workers Have Declined as the Industry Has Thrived." Center for Labor Research and Education. June. https://escholarship.org/uc/item/2t48t4sc.

Kurtz, Hilda, Abigail Borron, Jerry Shannon, and Alexis Weaver. 2019. "Community Food Assistance, Informal Social Networks, and the Labor of Care." *Agriculture and Human Values* 36 (3): 95–505.

Lathrop, Yannet, T. William Lester, and Matthew Wilson. 2021. "Quantifying the Impact of the Fight for $15: $150 Billion in Raises for 26 Million Workers, with $76 Billion Going to Workers of Color." Policy and Data Brief. National Employment Law Project. https://www.nelp.org/publication/quantifying-the-impact-of-the-fight-for-15-150-billion-in-raises-for-26-million-workers-with-76-billion-going-to-workers-of-color/.

Levine, Susan, and Steve Striffler, eds. 2015. "From Field to Table in Labor History." *Labor* 12 (1–2): 3–12. https://doi.org/10.1215/15476715-2837448.

Levinson, Marc. 2019. *The Great A&P and the Struggle for Small Business in America*. 2nd ed. Marc Levinson.

Liu, Yvonne, and Dominique Apollon. 2011. "The Color of Food." Applied Research Center. February. https://foodchainworkers.org/wp-content/uploads/2011/05/Color-of-Food_021611_F.pdf

Mares, Teresa. 2017. "Navigating Gendered Labor and Local Food: A Tale of Working Mothers in Vermont." *Food and Foodways* 25 (3): 177–192. https://doi.org/10.1080/07409710.2017.1343064.

Martin, Philip. 2003. "Promise Unfulfilled." Cornell University Press. https://www.cornellpress.cornell.edu/book/9780801488757/promise-unfulfilled/.

McCallum, Jamie K. 2022. *Essential: How the Pandemic Transformed the Long Fight for Worker Justice*. Basic Books.

McCarthy, Justin. 2022. "U.S. Approval of Labor Unions at Highest Point Since 1965." Gallup August 30. https://news.gallup.com/poll/398303/approval-labor-unions-highest-point-1965.aspx.

McNicholas, Celine, and Margaret Poydock. 2020. "Who Are Essential Workers? A Comprehensive Look at Their Wages, Demographics, and Unionization Rates." *Working Economics Blog*. May 19. https://www.epi.org/blog/who-are-essential-workers-a-comprehensive-look-at-their-wages-demographics-and-unionization-rates/.

Milkman, Ruth. 2022a. "The Amazon Labor Union's Historic Breakthrough." *Dissent Magazine*. https://www.dissentmagazine.org/online_articles/the-amazon-labor-unions-historic-breakthrough.

Milkman, Ruth. 2022b. "Grocery Unions Under the Gun in New York City and the Nation." *New Labor Forum* 31 (2): 17–26. https://doi.org/10.1177/10957960221091482.

Minkoff-Zern, Laura Anne. 2017. "Farmworker-Led Food Movements Then and Now: United Farm Workers, the Coalition of Immokalee Workers, and the Potential for Farm Labor Justice." In *The New Food Activism*, edited by Alison Alkon and Julie Guthman, 157–178. University of California Press. http://www.scopus.com/inward/record.url?scp=85038075041&partnerID=8YFLogxK.

Minkoff-Zern, Laura-Anne, and Teresa M. Mares. 2025. *Will Work for Food: Labor Across the Food Chain*. University of California Press.

Nieves, Alexander. 2022. "Newsom Reverses Course and Signs Farmworker Bill Backed by Biden." *Politico*. September 28. https://www.politico.com/news/2022/09/28/newsom-signs-biden-backed-union-bill-00056562.

Orleck, Annelise. 2018. *"We Are All Fast-Food Workers Now": The Global Uprising Against Poverty Wages*. Illustrated edition. Beacon Press.

Pangborn, Nicole, and Christopher Rea. 2020. "Race, Gender, and New Essential Workers During COVID-19." *Political Economy*. April 16. https://digitalcommons.wcl.american.edu/wlpeconomy/2.

Ramos, Athena K., Abigail E. Lowe, Jocelyn J. Herstein, Shelly Schwedhelm, Kelly K. Dineen, and John J. Lowe. 2020. "Invisible No More: The Impact of COVID-19 on Essential Food Production Workers." *Journal of Agromedicine* 25 (4): 378–382. https://doi.org/10.1080/1059924X.2020.1814925.

Rosenblum, Jonathan. 2017. "Fight for 15$: Good Wins, but Where Did the Focus on Organizing Go?" *Labor Studies Journal* 42 (4): 387–393. https://doi.org/https://doi.org/10.1177/0160449X17742799.

Sachs, Carolyn, and Patricia Allen. 2007. "Women and Food Chains: The Gendered Politics of Food." *International Journal of Sociology of Agriculture and Food* 15 (1): 1–23. https://doi.org/10.48416/ijsaf.v15i1.424.

Sbicca, Joshua. 2018. *Food Justice Now! Deepening the Roots of Social Struggle*. University of Minnesota Press.

Shreck, Aimee, Christy Getz, and Gail Feenstra. 2006. "Social Sustainability, Farm Labor, and Organic Agriculture: Findings from an Exploratory Analysis." *Agriculture and Human Values* 23 (4): 439–449. https://doi.org/10.1007/s10460-006-9016-2.

Walker, Richard. 2004. *The Conquest of Bread: 150 Years of Agribusiness in California*. New Press.

17 Creating Labor Sovereignty: Worker-Driven Movements for Justice

Vera Chang

> Some of the poorest, least powerful workers in this country—farmworkers divided by language, nationality and ethnicity, immigrant workers in a country itself deeply divided about their presence and contribution here—managed somehow over the course of two decades of struggle to forge a new form of worker power, a new paradigm for protecting vulnerable workers' fundamental human rights at the bottom of global supply chains.
>
> —Coalition of Immokalee Workers (2019)

National and multinational corporations devise and perpetuate systematic exploitation worldwide. This exploitation includes the violation of vulnerable peoples' right to feed themselves in just and self-determined ways. In response, Indigenous and peasant farmers have organized to demand *food sovereignty*—the "rights of peoples to healthy and culturally appropriate food produced through ecologically sound and sustainable methods, and their right to define their own food and agriculture systems" (Nyéléni International Movement for Food Sovereignty 2007). As inequalities within and beyond the food system have dispossessed farmers of their land and pressured them to migrate and become marginalized farmworkers, they have created new dreams of sovereignty.

One radical, liberatory vision is *labor sovereignty*—workers' right to define and ensure safe, healthy workplace conditions for themselves. Labor sovereignty positions workers as knowledgeable agents who can decolonize corporate-controlled structures to create emancipatory human rights–based supply chain regulations. This perspective shifts our understanding of labor justice as coming from above—as rights conferred by state institutions and corporate actors—to understand rights as defined by frontline workers with firsthand experiences of exploitation. Labor sovereignty therefore legitimizes and elevates worker expertise and respects their changemaking capacity.

This paper draws on my seven years of research as a second-generation Chinese American, feminist scholar-activist working with migrant tomato pickers and dairy workers mainly from Guatemala and Mexico as my research partners. Together, we have challenged notions of authority and belonging in the US food system, the normalization of violence against workers, and the stereotype that migrants are powerless. I am committed to this action-oriented, collaborative scholarship because my participation in social movements has taught me that bottom-up, cross-group (e.g., culture, citizenship status, language) alliances are necessary for groundbreaking, transformative food justice gains.

The Call for Labor Sovereignty

The labor sovereignty movement aligns with recent scholarship and activism that pushes us to think beyond the limits of dominant social, political, and economic systems to imagine and create radical conceptions of fairness and change (cf. Kojola and Pellow 2020). A vision of labor sovereignty rooted in workers' collective actions and social transformations depends on what Paolo Freire (2005 [1970]) calls "critical consciousness," or the ability of oppressed people to recognize and intervene in inequitable systems to improve their lives. Thus, social movements led by workers with little power can nevertheless create widespread transformation of working conditions.

Food and Multiple Sovereignties

Labor sovereignty emerges from the food sovereignty movement and its radical, feminist conceptualization of the territories, economies, and power that underpin the corporate food economy. Originally proposed by La Via Campesina in 1996, food sovereignty calls for deep and militant changes in production systems, including the democratization of power; transformation of political-economic foundations; return of rights, resources, and decision-making autonomy to peasants; and ontological transformation of the meaning of capital itself (Mares and Alkon 2011; Trauger 2014; Wittman 2011). This grassroots movement, known for its fight for communities' rights to land, seeds, and food, transcends local boundaries to decry the impacts of industrialization on people everywhere. This is critical in light of the widespread seizure of sovereignty from the state by private interests, which increases those private interests' already vast corporate regulatory power (Mares and Alkon 2011).

The 2007 Declaration of Nyéléni, written by over five hundred Indigenous, landless, worker, and other community representatives from eighty countries, identifies no single path to food sovereignty. Rather, it is a fluid and nuanced process (Wittman 2011). Sachs and Patel-Campillo (2014) describe food sovereignty as "another world feminism," which imagines an unconstrained, radically different world embodying multiple visions and demands to address inequality (Mares and Alkon 2011; Patel 2012). This is consistent with Iles et al. (2017), who argue that sovereignty's multiple forms enable social movements to construct and negotiate new kinds of power and pathways outside corporate control. Roman-Alcalá (chapter 1 in this book), Coté (chapter 5), and Ramirez (chapter 8) provide further discussion of struggles for decolonization and food sovereignty. Building on these critical food scholars, I suggest that the expansion of food sovereignty includes the governance practices of workers at the bottom of the industrial food chain.

Broadening our conception of food sovereignty, I draw on Bartra's (2004) notion of labor sovereignty, which describes how the North American Free Trade Agreement (NAFTA) dispossessed Mexican peasants of rural lands, denying their food sovereignty and forcing them to emigrate. In this context, labor sovereignty is a nation's ability to provide living wages to a majority of the population. Building on Bartra, Otero (2011) suggests that the achievement of labor sovereignty will require agricultural land-based reforms attentive to worker rights for Mexican peasants.

However, Otero cautions that even if Mexican peasants achieve labor sovereignty, "those concerned with human and labor rights will [still] have to address [NAFTA's] effects on migrants in the US" (Otero 2011, 386). Struggling to sustain themselves, many former peasants in Mexico, across Latin America, and worldwide have had to move to industrialized nations as undocumented migrant workers, reliant on wage labor instead of working their own land. Separated from their families, their homes, and the products of their labor, they have joined a vastly different form of agriculture in which massive scale and profit motives intensify heavy exploitation of people, land, and animals. (See chapter 16 of this book by Mares and Minkoff-Zern on the normalization of exploitation and the need to center worker justice in the food system.)

Thus, in the neoliberal state, corporations at one end of the power spectrum are seen as both the unquestioned experts on and enforcers of supply chain conditions. At the other end, workers carry the modern agricultural system on their backs. As Mares and Alkon (2011) observe, industrial food production is built on systematic and often violent denial of sovereignty to workers throughout the food chain (see also Mares 2019). This dichotomy is particularly striking in the US food system, where concentrated corporate power excludes farmworkers from most legal protections, while the few protections that these workers have go unmonitored and unenforced. Indeed, in the US, migrant farmworkers are among the nation's most marginalized and lowest-paid workers in one of the most dangerous occupations (Bon Appétit and United Farm Workers 2011). The vast labor inequalities within the current market economy demand a second conception of labor sovereignty.

A New Definition of Labor Sovereignty

Working within this political-economic environment and responding to Bartra (2004) and Otero (2011), I propose a new definition of labor sovereignty that is attentive to migrant workers' struggles for control of their workplace conditions in corporate agrifood chains. Despite the forces stacked against them, some worker groups have begun transforming vast industrial systems by demanding labor sovereignty. Rooted in established human and democratic rights, labor sovereignty is the right of workers to define, demand, and oversee safe, healthy, and dignified working and living conditions for themselves. By centering workers' expertise and leadership, labor sovereignty repositions subjugated workers within new sociocultural knowledge systems that respect their know-how and their rights. This definition extends scholarly understandings of both food and labor sovereignty to include workers in contexts where capital and labor are differentiated entities and where workers absorb the environmental and social contaminants that are perpetually externalized in capitalist agricultural systems (cf. Wittman's 2011 discussion of the "metabolic rift").

The case study I describe in the following section reveals that an emergent form of organizing, *worker-driven social responsibility,* enacts labor sovereignty in unprecedented ways using innovative governance tactics.

Labor Sovereignty in Practice: Worker-Driven Social Responsibility in US Agriculture

Worker-driven social responsibility (WSR) is a human and labor rights framework created by workers to challenge the exploitation, poverty, and injury ubiquitous in industrial supply chains. Coined by a Florida-based tomato worker–led human rights group, the Coalition of Immokalee Workers (CIW), WSR enables workers in disparate industries to institute legally binding programs to enforce their rights.

The first comprehensive WSR program, the CIW's Fair Food Program, arose from a decades-long struggle by some of the poorest, least powerful communities in the US—Latinx, Mayan, and Haitian immigrants in the agricultural fields of Florida, dubbed "ground zero for modern [US] slavery" (Bowe 2003). The seeds of the Fair Food Program date back to the CIW's 2001 Taco Bell boycott, which called on the burrito chain to pay a penny more per pound for tomatoes to address workers' subpoverty wages and to buy exclusively from growers who respect workers' rights (CIW 2005). In 2005, this led to the first legally binding agreement between workers at the bottom of a corporate supply chain and a brand at the top, establishing the fundamental elements of the WSR model. After a ten-year campaign resulting in binding agreements with nine major retail and fast-food companies, the CIW accumulated enough buying power to launch the Fair Food Program in 2011. Today, the Fair Food Program spans crops from peppers to flowers across ten US states, Chile, and South Africa through agreements with fourteen major food and crop buyers.

Contemporaneously, outside the food system, another worker-led effort to hold big business accountable was evolving in Bangladesh's garment industry, the site of ghastly factory fires and building collapses, where workers and labor advocates had been fighting for lifesaving safety measures. In 2011, inspired in part by binding university labor codes secured by the student antisweatshop movement, earlier jobber agreements in the US apparel industry, and the CIW's breakthrough agreements, a consortium of unions and labor rights organizations began constructing the Bangladesh Accord, launched in 2013 after the deadly Rana Plaza building collapse that killed over a thousand workers. Today, what is now known as the International Accord (the Accord) protects three million workers across two thousand

Bangladeshi and Pakistani factories with over two hundred apparel brand signatories.

Together, using structural blueprints from the Fair Food Program and Accord, the WSR model continues to expand as a global movement into diverse industries, locales, and economic contexts facilitated through the Worker-driven Social Responsibility Network, founded in 2015. The first major adaptation of the WSR model in the US was implemented in 2018 by Migrant Justice, a Vermont-based, dairy worker–led human rights group struggling against worker fatalities and economic injustice through its Milk with Dignity program in global ice cream maker Ben & Jerry's supply chain (Chang 2018). Local worker groups also implemented WSR in Lesotho's garment industry and are exploring its implementation in Minnesota's construction industry and Arkansas's poultry industry.

In the rest of this chapter, I draw on my research on the Fair Food Program and the Milk with Dignity program to examine how farmworkers assert themselves as supply chain experts and leaders in corporate-controlled industries, thereby championing new conceptualizations of fairness, justice, and power.

How Worker-Driven Social Responsibility Works

WSR's bottom-up model enables workers to lead a social movement using a market-based system that strictly enforces worker-defined rights. In the tradition of farm labor union advocacy and other worker organizing that preceded it, WSR groups creatively and strategically engage with legal categories and norms while enlisting allies to join them in building popular movements that bring pressure to bear on powerful decision-makers (Dias-Abey 2018). The need for creativity emerges from the legal reality that racialized farmworkers are explicitly excluded from the federal right to unionize—and thus lack access to the essential change-making tactic of collective bargaining. When and where unions are allowed, they are subject to legal constraints (e.g., no secondary boycotts), and it is risky for undocumented workers to engage in worker-led protest. These are just a few of the reasons that farm labor is among the least unionized industries in the US (Mayer 2004) and why farmworker unions, like the United Farm Workers (UFW), have encountered major hurdles when seeking to advance workers' rights over the years.

Florida-based farmworker advocates with the CIW have responded to these challenges by creating new forms of activism that draw on both successful union organizing techniques and alternative legal strategies to protect workers' rights. UFW president Arturo Rodriguez called CIW's first major boycott victory "the most significant . . . since the successful grape boycott led by the UFW in the 1960s in the fields of California" (Leary 2005). Unlike unions, in which workers leverage their legal rights via court systems to bargain directly with their employers, WSR workers hold market power. They can bargain up the supply chain to influence and enforce conditions of employment by targeting suppliers' commercial interests. Thus, WSR's political-economic approach overcomes the limitations of public regulation by allowing workers to vindicate rights regardless of migration status, counterbalancing migrants' "unfreedom" (Dias-Abey 2018, 207). In doing so, WSR musters political, economic, and cultural resources beyond those of unions.

WSR also differs radically from conventional private approaches to change-making like consumer-based corporate social responsibility and multistakeholder initiatives (Minkoff-Zern 2017). A growing body of literature exposes how these mainstream efforts to further sustainability and human rights have failed workers. In contrast to corporate-, multistakeholder-, and consumer-led campaigns in which change-making is constrained to individuals "voting with their dollars," WSR campaigns are part of a broader movement for consumer-farmworker solidarity. Beyond enlisting individual consumer purchasing power, WSR utilizes the full breadth of consumer power by encouraging allies to engage in boycotts, protests, and media campaigns (Minkoff-Zern 2017).

WSR's mix of enforcement mechanisms allows workers to control their own human rights in the following ways:

- Creating a WSR program begins with community assemblies where workers enumerate the issues their program should cover. Workers coauthor codes of conduct for farms based on firsthand experience. Examples include zero tolerance for forced labor, child labor, and sexual assault conditions; pay for all hours worked; and mandatory cool-down rest breaks. Workers continually revise standards based on their own experiences and those voiced by their peers during education sessions, audits, grievance filings, and community meetings. The most marginalized workers, such as women workers, lead the creation of WSR standards.
- Worker-led, consumer ally–backed campaigns pressure retail and fast-food chains to join industry-specific WSR programs. Buyers who agree to join provide direct support to workers through bonuses paid as a premium and

fund improvements to workplace conditions. Participating corporations also preferentially source from suppliers who consent to uphold workers' standards.

- Worker-to-worker education sessions train workers to defend their rights through continuous monitoring, such as by filing complaints through a 24/7 confidential hotline staffed by a third-party monitoring body specific to each WSR program (e.g., Fair Food Standards Council, Milk with Dignity Standards Council). Overseen by the worker organization, these monitoring bodies implement, monitor, and enforce the programs, which include conducting complaint investigations and ongoing audits, issuing corrective action plans for farms, and responding to worker grievances. (See Chang 2020 for a discussion of the ways that WSR monitoring is legal-like and can be more effective than the judicial system.)
- Backed by the purchasing power of multi-million-dollar brands, each WSR program vigilantly protects workers who speak out after retaliation and other violations via immediate, legally binding economic consequences for farms (e.g., withholding corporate sales). These market-based sanctions effectively incentivize suppliers to cooperate with workers' standards (Chang 2020).
- Ultimately, each WSR program is accountable to a workers' council, ensuring that workers' interests are not subverted in the broader supply chain structure where corporate power predominates.

WSR's unique combination of tactics has effectively advanced workers' demands, successfully remediating and preventing exploitation in the workplace. This is evident in published program results demonstrating workers' gains in enforcing their rights through WSR (Fair Food Standards Council 2021; Milk with Dignity Standards Council 2022). In particular, patterns of complaint data elucidate working conditions, workers' confidence in the system, and the balance of power between workers and bosses. Meanwhile, compliance rates further evince the programs' validity, illustrating that the Fair Food Program and Milk with Dignity are transforming workplace systems like hiring, hours, and safety structures.

The first example of WSR, the Fair Food Program, has resolved over three thousand grievances—roughly one complaint per day since the program's inception—with 65 percent of these remedied in under two weeks. Since 2011, the program has suspended seven growers and placed twenty-four on probation. Nearly all suspensions took place in the Fair Food Program's first few years, when compliance also improved most drastically (Fair Food Standards Council 2021).

These WSR successes have spread. For instance, minimum wage compliance jumped from 60 to 90 percent in Milk with Dignity's first five years of operation (Milk with Dignity Standards Council 2022). Since 2018, Milk with Dignity has resolved nearly five hundred code violations through its 24/7 support line alone. The result has been near-complete compliance, with workers having at least eight consecutive hours off each day, paid sick leave, and access to a bedroom with a door—rights previously unheard of in the dairy business. Among Milk with Dignity's numerous achievements has been an upward trend enforcing the provision of protective equipment for handling dangerous chemicals like formaldehyde and copper sulfate. This is not inconsequential in an industry where more than four in five dairy workers report harm from chemical and biological exposure (Milk with Dignity Standards Council 2022). Vermont farmworker Enrique Balcazar testified that Migrant Justice chose to replicate the WSR model in the dairy industry because it has "real teeth, real enforcement, and real worker participation" (Chang 2015). (See figure 17.1 for an example of a participating Milk with Dignity farm.)

These changes challenge the deeply rooted assumption that farmworkers must endure unsafe conditions. Because of WSR's effectiveness in advancing workers' demands, the Equal Employment Opportunity Commission (2016) called WSR a "radically different accountability mechanism," while the MacArthur Foundation (2017) called it a "visionary strategy . . . with potential to transform workplace environments across the global supply chain." Due to the resources and power required to force corporate hegemony to cede any control to workers, few movements that struggle for labor sovereignty actually achieve it. WSR provides an inspiring blueprint for developing effective, rights-affirming processes transferable to other regions, industries, and economic contexts.

Democratic Worker-Led Human Rights Governance

As a democratic, bottom-up epistemology, WSR is rooted in the leadership of workers with firsthand experience of exploitative, inequitable labor arrangements. Unlike corporate social responsibility and multistakeholder initiatives that tout "stakeholder participation" in their governance schemes without meaningfully including workers (MSI Integrity 2020), WSR is led by workers

Figure 17.1
A worker in the barn of a Vermont dairy trained to monitor his rights under Migrant Justice's Milk with Dignity program, modeled after the Coalition of Immokalee Workers' Fair Food Program, which provides a blueprint for the WSR model. Photo by Vera Chang.

and harnesses their broad participation. In addition, unlike collective bargaining strategies, WSR workers' governance strategies are unconstrained by the same legal restrictions as traditional labor unions. Leading their own programs for human and labor rights, workers challenge the naturalized rules and hierarchical order of the corporate supply chain, building new, democratic labor knowledge systems that value workers' demands for rights and fairness. Luis C.deBaca, former ambassador-at-large to monitor and combat trafficking under President Barack Obama, worked with the Coalition of Immokalee Workers in the 1990s to tackle some of the highest-profile modern slavery cases in the US, changing the landscape of the antitrafficking movement. C.deBaca noted how extraordinary it is for workers to regulate their own conditions: "The premise that workers should design and control their own program for remedying and monitoring their own labor and human rights is unusual—even revolutionary—because it is rarely, if ever, done" (quoted in Chang 2020).

The corporate food regime does not position marginalized workers as "knowledge generators" (Iles et al. 2017, 957). In contrast, WSR's governance structures are grounded in what I call "worker-made knowledge." Drawing from Iles et al.'s (2017) conception of farmer-made knowledge, worker-made knowledge demonstrates how WSR allows workers to draw on their own skilled visions (Grasseni 2005) and expert knowledge (Fischer 2000) to build new social-cultural systems. This new regulatory order of agricultural labor values workers' expertise, needs, and demands for justice. Iles et al. (2017) observe that the coproduction of knowledge and labor can create countercurrents to the dominant system that could transform industrial agriculture's epistemic underpinnings.

Workers' struggles to legitimize their knowledge and agency align with Freire's (2005 [1970]) conception of critical consciousness. By developing workers' collective awareness, WSR helps workers free themselves from a culture of silence to instead engage in what Freire (2005 [1970], 34) calls the "practice of freedom" from oppressive structures: "the means by which men and women deal critically and creatively with reality and discover how to participate in the transformation of their world."

Intersectional Bottom-Up Worker Justice Activism

Rooted in the successes of resonant, powerful sovereignty movements that came before it—such as Haitian, Latin American, and US civil rights organizing—the Fair Food Program and Milk with Dignity demonstrate how workers can have a radical voice in established production systems that have excluded them. To build the movement, worker-directed campaigns pressure corporations to join WSR programs, drawing on consumer solidarity to apply influence through public actions (Minkoff-Zern 2017).

This rights-based check on power seeks to address the complexity of corporate-controlled labor systems along multiple dimensions. Iles and Montenegro de Wit (2015) aver that multiscalar food systems draw power from different locales (i.e., institutions, allies, places) capable of redistributing knowledge and resources quickly and nonhierarchically. Applying this idea, we see that WSR's multiscalar approach makes progress toward disrupting capital's velocity, wielding new kinds of power, and altering the structures and processes that underlie the corporate food regime. At the farm level and aligned with Freire's (2005 [1970]) popular education for critical consciousness, workers use

WSR to teach each other about structural issues, their rights, and collective action tactics to gain power. This is similar to farmer-to-farmer popular education practices that have spread throughout the global food sovereignty movement. At grassroots, national, and international scales, WSR's work extends through the Worker-driven Social Responsibility Network, which emphasizes workers' local and technical knowledge in diverse sectors, geographies, and sociopolitical contexts.

To gain legitimacy and achieve sovereignty over their working conditions, WSR-monitored workplaces utilize third-party monitoring bodies (e.g., Fair Food Standards Council, Milk with Dignity Standards Council) that provide workers with mechanisms to claim their rights under their community-written code of conduct. WSR programs further institutionalize workers' authority over supply chain governance through partnerships with state institutions. For example, WSR organizations have worked with federal prosecutors, the Department of Labor, and other law enforcement to successfully prosecute human trafficking, forced labor, assault, and other illegal conduct. This has ended in referrals to the Federal Bureau of Investigation and Department of Homeland Security for criminal investigation and long prison sentences for abusive bosses and their acolytes (CIW 2013), holding industry accountable for practices in their supply chains. Amengual and Fine (2017) argue that "co-enforcement" between worker-run private governance programs and public regulation can effectively enforce labor rights.

In the presence of governmental unwillingness or political inability to institute strong regulations, WSR's tactics help workers gain control and assert their interests in our market-based economy. Coupling strictly enforced, worker-authored private governance standards with a social movement, WSR makes creative use of capitalistic tools, compelling brands to adopt strong community standards and codify them in their corporate operating procedures. WSR's heterogeneous approach exemplifies what Alkon and Guthman (2017) call "intersectional food activism," which works within the logic of the current neoliberal political economy while pushing beyond it to address hierarchies of inequality (see also Trauger 2014). My research reveals that WSR's multidimensional processes enable workers to make powerful demands to reclaim and enforce their self-defined rights within neoliberal capitalism. It is this multiscalar combination of strategies that makes WSR so radical and effective, enabling workers to win back their sovereignty from corporations (Chang 2020).

Formation of Counterhegemonic Worker Consciousness

Donella Meadows's (1999) idea of "leverage points" helps us conceptualize how systems can change. Meadows explains how small-scale actions in complex systems, like corporate food supply chains, can produce fundamental systemic changes—and thus more sovereign possibilities. According to Meadows, there are always multiple potential interventions for change. Meadows's theory is consistent with the way that Trauger (2017) and Mares (2019) describe food sovereignty as embodied in "everyday acts of defiance" that can ultimately generate major shifts. Effective labor sovereignty practices lead to the creation of alternative labor paradigms and subvert accepted beliefs about the efficacy and inevitability of existing hierarchies.

This effectiveness is evident in WSR processes through which workers have gained authority within corporate supply chain territories. Thousands of workers' participation in multiple, overlapping proactive and reactive enforcement mechanisms designed to achieve compliance with worker-defined rights has redistributed rights and wages and improved workers' health and safety, reshaping the workplace environment (Fair Food Standards Council 2021; Milk with Dignity Standards Council 2022). Seemingly minor remedies, like returning stolen wages or preventing verbal harassment, can add up to transform entrenched food production chain practices, norms, and relations. While there will always be more work to do, implementing WSR standards means eradicating discrimination, harassment, and assault from participating farms and supply chains. In the process of doing so, the global WSR movement has catalyzed an effective process for gaining power for workers.

Conclusion

This chapter describes how bold worker imaginaries of labor sovereignty can challenge and transform commercial agriculture's workplace norms in the face of persistent corporate hegemony. It explains how abusive workplace practices can shift when impunity is met with enforced accountability to reduce exploitation and foster justice. By theorizing how labor sovereignty can take root in the US food system, this chapter brings the food sovereignty movement full circle, tracing the development of resistance struggles by disenfranchised people who have migrated to find work for their survival.

As a practice of labor sovereignty, WSR imagines work beyond the limits of current systems to envision how reorganizing supply chain labor relations according to workers' vision can improve work and lives. At the forefront of creating a more powerful, effective food politics, the WSR movement exemplifies how sovereignty struggles can develop, evolve, and find victory while modeling a more ethical, worker-centered food system. Understanding the process of bottom-up systemic change—what it is, how to achieve it, and how it can influence contemporary food labor systems—is essential. As workers take on their rightful roles as labor, human rights, and movement leaders, we will see increasingly radical transformations toward more just industrial labor systems.

References

Alkon, Alison, and Julie Guthman, eds. 2017. *The New Food Activism: Opposition, Cooperation, and Collective Action*. University of California Press.

Amengual, Matthew, and Janice Fine. 2017. "Co-enforcing labor standards: The unique contributions of state and worker organizations in Argentina and the US." *Regulation & Governance* 11 (2): 129–142. https://doi.org/10.1111/rego.12122.

Bartra, Armando. 2004. "Rebellious cornfields: Toward food and labour self-sufficiency." In *Mexico in Transition: Neoliberal Globalism, The State, and Civil Society*, ed. Gerardo Otero, 18–36. Zed Books.

Bon Appétit Management Company and United Farm Workers. 2011. Inventory of Farmworker Issues and Protections in the US. https://www.bamco.com/wp-content/uploads/2016/08/farmworkerinventory_0428_2011_updated2016.pdf.

Bowe, John. 2009. "Nobodies: Does slavery exist in America?" *New Yorker*, April 13. https://www.newyorker.com/magazine/2003/04/21/nobodies.

Chang, Vera. 2015. "Will Ben & Jerry's help improve conditions for dairy workers?" Civil Eats. https://civileats.com/2015/05/11/will-ben-and-jerrys-help-improve-conditions-for-dairy-workers/.

Chang, Vera. 2018. "Disruptions in Vermont's landscape." Social Science Research Council. https://www.racheltanurmemorialprize.org/disruptions-in-vermonts-landscape/.

Chang, Vera. 2020. "After #MeToo, this group has nearly erased sexual harassment in farm fields." Civil Eats. https://civileats.com/2020/03/09/after-metoo-this-group-has-nearly-erased-sexual-harassment-in-farm-fields/.

Coalition of Immokalee Workers. 2005. "Taco Bell agreement analysis." https://ciw-online.org/blog/2005/03/agreement-analysis/.

Coalition of Immokalee Workers. 2013. "Breaking: US Department of Labor press release on latest forced labor prosecution names buyers, including Fair Food Program holdout Kroger." https://ciw-online.org/blog/2023/02/breaking.

Coalition of Immokalee Workers. 2019. "OP/ED: An astonishing journey from Lesotho to Immokalee and back. https://ciw-online.org/blog/2019/10/op-ed-lesotho-albisa/.

Dias-Abey, Manoj. 2018. "Justice on our fields: Can 'alt-labor' organizations improve migrant farm workers' conditions?" *Harvard Civil Rights-Civil Liberties Law Review* 53 (1): 167–211.

Equal Employment Opportunity Commission. 2016. "Select Task Force on the Study of Harassment in the Workplace." https://www.eeoc.gov/select-task-force-study-harassment-workplace.

Fair Food Standards Council. 2021. "Fair Food Program 2021." https://fairfoodprogram.org/results/.

Fischer, Frank. 2000. *Citizens, Experts, and the Environment: The Politics of Local Knowledge*. Duke University Press.

Freire, Paulo. 2005 [1970]. *Pedagogy of the Oppressed*. Continuum.

Grasseni, Cristina. 2005. "Designer cows: The practice of cattle breeding between skill and standardization." *Society & Animals* 13 (1): 33–50. http://doi.org/10.1163/1568530053966652.

Iles, Alastair, Garrett Graddy-Lovelace, Maywa Montenegro de Wit, and Ryan Galt. 2017. "Agricultural systems: Co-producing knowledge and food." In the *Handbook of Science and Technology Studies*, 4th ed., ed. U. Felt, R. Fouché, C. Miller, and L. Smith-Doerr, 943–972. MIT Press.

Iles, Alastair, and Maywa Montenegro de Wit. 2015. "Sovereignty at what scale? An inquiry into multiple dimensions of food sovereignty." *Globalizations* 12 (4): 481–497. https://doi.org/10.1080/14747731.2014.957587.

Kojola, Erik, and David Pellow. 2020. "New directions in environmental justice studies: examining the state and violence." *Environmental Politics* 30 (1–2): 100–118. https://doi.org/10.1080/09644016.2020.1836898.

Leary, Elly. 2005. "Immokalee workers take down Taco Bell." *Monthly Review*, October 1, 11–25. https://monthlyreview.org/2005/10/01/immokalee-workers-take-down-taco-bell/.

MacArthur Foundation. 2017. "Greg Asbed: Transforming conditions for low-wage workers with a visionary model of worker-driven social responsibility." https://www.macfound.org/fellows/class-of-2017/greg-asbed.

Mares, Teresa. 2019. *Life On the Other Border: Farmworkers and Food Justice in Vermont.* University of California Press.

Mares, Teresa, and Alison Alkon. 2011. "Mapping the food movement: Addressing inequality and neoliberalism." *Environment and Society* 2 (1): 68–86. https://doi.org/10.3167/ares.2011.020105.

Mayer, Gerald. 2004. "Union membership trends in the United States." Congressional Research Service. http://digitalcommons.ilr.cornell.edu/key_workplace/174.

Meadows, Donella. 1999. *Leverage Points: Places to Intervene in a System.* The Sustainability Institute.

Minkoff-Zern, Laura Anne. 2017. "Farmworker-led food movements then and now: United Farm Workers, the Coalition of Immokalee Workers, and the potential for farm labor justice." In *The New Food Activism*, ed. Alison Alkon and Julie Guthman, 157–178. University of California Press.

Milk with Dignity Standards Council. 2022. "2018–2022: Five Years of Milk with Dignity." https://milkwithdignity.org/sites/default/files/2022_md_program_five-year_report.pdf.

MSI Integrity. 2020. "Not fit-for-purpose: The grand experiment of multi-stakeholder initiatives in corporate accountability, human rights, and global governance." https://www.msi-integrity.org/not-fit-for-purpose/.

Nyéléni International Movement for Food Sovereignty. 2007. "Declaration of Nyéléni 2007." https://nyeleni.org/en/declaration-of-nyeleni/.

Otero, Gerardo. 2011. "Neoliberal globalization, NAFTA, and migration: Mexico's loss of food and labor sovereignty." *Journal of Poverty* 15 (4): 384–402. https://doi.org/10.1080/10875549.2011.614514.

Patel, Raj. 2012. *Stuffed and Starved: The Hidden Battle for the World Food System.* 2nd ed. Melville House.

Sachs, Carolyn, and Anouk Patel-Campillo. 2014. "Feminist Food Justice: Crafting a New Vision." *Feminist Studies* 40 (2): 396–410. https://dx.doi.org/10.1353/fem.2014.0008.

Trauger, Amy. 2014. "Toward a political geography of food sovereignty: Transforming territory, exchange and power in the liberal sovereign state." *Journal of Peasant Studies* 41 (6): 1131–1152. https://doi.org/10.1080/03066150.2014.937339.

Trauger, Amy. 2017. *We Want Land to Live: Making Political Space for Food Sovereignty.* University of Georgia Press.

Wittman, Hannah. 2011. "Food Sovereignty: A new rights framework for food and nature?" *Environment and Society* 2 (1). https://doi.org/10.3167/ares.2011.020106.

18 First Food Equity: A Reparative Approach

Chellamal Keshavan

I was a twenty-three-year-old student working hard in pursuit of my master's in public health when I chose to carry a pregnancy to term for the first time. At that age, I naively thought I had all the tools necessary to embrace motherhood. I had earned my bachelor's in child development and had worked in Reggio-Emilia preschools, afterschool programs, and camps, not to mention years of babysitting. Furthermore, I had been raised with a South Indian reverence for the identity of motherhood. In fact, my name, Chellamal, means "sweetness of motherhood" in Tamil, one of the several South Indian Dravidian languages. After learning about what Harvey Karp, pediatrician and child development expert, called "the fourth trimester," defined as the first twelve weeks after birth, I felt that I had found my calling (Karp n.d.). I have always been my best self in the zero-to-three world, finding more joy and ease with our youngest community. But despite my significant theoretical and practical education in early childhood, I found myself ill-prepared for the emotional trajectory of motherhood and surrounded by adults wielding a combination of seniority and inexperience in the fourth trimester.

Nine months of pregnancy and thirty-six hours of unmedicated labor made me a parent for the first time. I met my new baby with a combination of love, amazement, bewilderment, and fear. Soon after his delivery, an exhaustion that I had never experienced before set in; my whole body hurt, and yet the work of navigating the first latch was ahead of me.

Adoption and youth coalesced to create an emotional isolation; as a transnational adoptee, I did not have parents who had experienced pregnancy or childbirth themselves, and because I was the first of my peer group to have a baby, I did not have the support of experienced women around me. When

I learned that every female baby is born with all the eggs that she will own in her lifetime, I leaned into the comfort that my birth mother had carried the potential that has become both my children, *that she carried all three of us* (Henigsman n.d.). The process of becoming a mother left me feeling failed by both mothers: one who birthed me without parenting and the second who parented without birthing me. It is the connection with the birth mother that helped me to feel prepared for the work of birth, breastfeeding, and mothering.

I also remember feeling betrayed by the copious texts I read describing breastfeeding with gentle language like "natural" and "feminine." I would largely breastfeed in partnership with my babies with little perceived support from my surrounding community. Despite these barriers, I would go on to nurse my first child for two years and my second for two years and three months. The experience of breastfeeding felt like an emotional marathon. A pendulum shifting in perpetuity from oxytocin-informed feelings of deep love and attachment to feelings of resentment over the amount of work breastfeeding requires, often expressed through jealousy over the freedom of family members and friends who were not nursing. To be clear, exclusive breastfeeding requires about 25 percent of the body's energy and about 650 kcal per day. By comparison, the brain runs on a modest 20 percent of the body's energy (Porter n.d.). And yet, I came to understand breastfeeding as an incredible symbiosis that holds its own relationship with time, suspending my existing understanding of productivity and value in support of centering the benefits of breastmilk and attachment afforded to the parent-baby dyad. Breastfeeding became a process where beauty and challenge coalesced to create two new people. Every time a new baby is born, a new parent is born.

While exclusive breastfeeding was the right choice for my baby and I, "fed is best" has become a personal mantra, as feeding practices that support familial efficacy and happiness are paramount. The beginning hours and days of my breastfeeding experience felt lonely, surprisingly physically painful, and time consuming. The memories of that time solidified respect for all of the ways that babies successfully feed depending on their parents' personal best practices. I would go on to breastfeed exclusively for six months, and in partnership with solids for a year and half after that. I learned intimately how physically, emotionally, and financially challenging exclusively breastfeeding is. Few parents have access to intersectional supports that inform best breastfeeding practices. It is necessary to honor what pregnancy-capable parents decide is correct for themselves. The experience of being physically and emotionally available on demand would add to my social conditioning a personal understanding of the extent to which our society fails parents and babies during the perinatal years.

The experience of having a baby biologically when my own mother became a mother by adoption would lead me to become a peer breastfeeding community transformer through Reaching Our Sisters Everywhere based in Atlanta and then a certified lactation specialist and postpartum doula through the Boston Breastfeeding Coalition and Doulas of North America, respectively. I was driven to become the support person that I needed while pregnant and didn't have access to. Individualized parent-baby work would become systems-based education, program design, and implementation when I stepped into a role as executive director at a maternal health equity nonprofit. In the last ten years, my experiences with breastfeeding have evolved from dyadic and personal into larger concentric circles, positioning breastfeeding equity as an international indicator of social progress.

For the purposes of this chapter, I will first explore the concept of "first food equity" in the US and the systemic issues of racialized breastfeeding inequalities. First food equity is the shared understanding that equitable support for breastfeeding is critical to the health and wellness of parents and babies who choose it. First food equity aims to create the conditions whereby parents of all backgrounds can realize their chosen breastfeeding goals. Equitable support includes intentional access to affordable and culturally responsive lactation education, community-based opportunities for learning in partnership with other breastfeeding families, evidence-based training and appropriate wages for breastfeeding practitioners who are in and of communities of color, and family-enabling policies, such as the Family and Medical Leave Act, that allow breastfeeding parents the financial safety net they need in order to focus on early parenthood.

In this chapter, I will propose strategies for potential avenues of partnership that stand to strengthen both food justice and first food equity moving forward. The food justice movement already holds the tools necessary to engage and support a breastfeeding justice agenda, and thus first food equity should be considered a critical component of the larger food justice movement. Given the staggering inequities in maternal health, first food equity should be informed by the ways that food justice advocacy centers and lifts voices that are farthest from opportunity, a deep respect for ethnographic and qualitative data, and the capacity to foreground a feminist lens to advance regional conversations that inform global capacity. Furthermore, first food equity can foreground the need for intersectional feminist approaches to food justice, which have too often been confined to the margins. This chapter will highlight some of the most cogent opportunities for alignment.

Food Justice from a Feminist Lens

Carolyn Sachs and Anouk Patel-Campillo (2014) foreground what they call "feminist food justice." As they explain, "Despite the clearly gendered dimensions of hunger and food insecurity, major approaches to tackling hunger, including food security efforts by the United Nations, food sovereignty movements in developing countries, and food justice movements in the United States, do not place gender issues at the center" (398). The sociopolitical exercise of engaging in conversation concerning first food equity could serve the food justice movement in exploration of how gender intersects with unequal food systems. Sachs and Patel-Campillo emphasize that "women's bodies are often sites of malnutrition" (398). Furthermore, Allen and Sachs (2007) explain, "Today in most societies women continue to carry the responsibility for the mental and manual labor of food provision—the most basic labor of care." Despite the most supportive allies, breastfeeding remains primarily the work of the birthing parent, and so a commitment to first food equity would require a feminist approach.

Sachs and Patel-Campillo (2014) have centered a deep understanding of international feminisms in their discussion of gender and food justice. They invite collaborative thinkers to consider the scope at which they understand intersectionality (404). In her paper, "From Patriarchy to Intersectionality: A Transnational Feminist Assessment of How Far We've Really Come," Vrushali Patel offers a new angle with which to understand globalization and food justice: "From the perspective of intersectionality, merely reframing these dynamics through the lens of domestic intersectionality will only perpetuate the reification of the local. We need to recenter the notion that there are no locals and globals, only locals in relation to various global processes" (2013, 847). Patel's perspective is particularly salient in the first food equity space, as it could help emphasize that breastfeeding can be considered and supported as a global capacity. The relationship between the local and global processes may support the design of relevant policy in supporting breastfeeding choices for pregnancy-capable populations from all backgrounds. Building on Sachs and Patel-Campillo's call to understand food justice from a perspective of intersectional feminisms, we can choose to understand first food equity through the lens of Black feminist thought.

In their article, "Making Spaces Something Like Freedom: Black Feminist Praxis in the Re/Imagining of the Just Food System," anthropologist Ashante Reese and activist Dara Cooper explore the evolution of the relationship between the food justice movement and racism over the last twenty years. Reese and Cooper detail the importance of developing a pedagogy of freedom from a Black feminist lens. To them, a Black feminist approach is "one that centers analyses of racialized sexisms and homophobia, and that foregrounds Black women as intellectual producers, as creative agents, as political subjects, and as freedom dreamers" (2021, 452). Attention to first food equity through the lens of reparations as described below, is one way to enact this approach.

Allen and Sachs (2007) situate the burden of food and feeding by explaining that "historically, women have been associated with domesticity and home spaces" including nourishment and food production. Reese and Cooper appropriately complicate an association of domesticity by reminding readers that the American Black woman has not consistently been afforded identity formed along traditional white middle-class gender roles. Reese and Cooper build on Sachs and Allen's commentary: "The distinctions between home and public spheres are complicated for Black women, because of threats of violence, being seen as ungendered laborers, and the need for multiple income streams to support a single household, which often required Black women to be in white women's homes more than their own" (2021, 452). The American Black woman may experience relationships with food as a spectrum between subjugation and freedom. Reese and Cooper offer a framework of relationality as an opportunity for healing, positing community care as a salient avenue toward solidifying a working relationship between food justice and multiple iterations of feminisms. For them, care is a collective responsibility: "Care as shared risk is not a theoretical construct absent of application. Instead, we understand it to be an overarching, aspirational and a relational framework that can be applied in a variety of organizational structures" (2021, 454). Breastfeeding is a form of care work in which the body of the birthing parent supports and cares for the infant. By proposing culturally informed lactation support as well as policy interventions, first food equity becomes a means for the food justice movement to realize the "care as shared risk" that Reese and Cooper envision.

Breastfeeding occupies an incongruent dichotomy in that it gleans so little structural respect and support while simultaneously receiving consistent and unsolicited feedback. Cultural expectations shape the extent to which breastfeeding parents feel accepted and supported. First food equity and, by extension, food justice have the opportunity to engage a feminist lens from a place of belonging. A dual mechanism that navigates a community-based celebration of parents and babies acting in conjunction with a policy informed arm that seeks to codisrupt existing boundaries to first food equity might inform a feminist angle that acknowledges spatial and racial relativity moving forward. In fact, first food equity may be useful in exploring a larger conversation about racialized reparations.

The Case for Breastfeeding Reparations

The National African-American Reparations Commission (NAARC), convened and administered by Black World 21st Century, is a group of distinguished and highly achieving leaders from academia, law, and professional backgrounds. The NAARC (n.d.) defines its mission as being "united in a common commitment to fight for reparatory justice, compensation and restoration of African-American communities that were plundered by the historical crimes of slavery, segregation and colonialism and that continue to be victimized by the legacies of slavery and American apartheid." Furthermore, the Commission has authored a ten-point reparation plan encompassing intersectional issues of systemic racism and white supremacy. Section five of the plan is titled "Health and Wellness," and explores issues of healthy equity. Section 5A calls for "funds for the establishment of a regional system of Black controlled Health and Wellness Centers, fully equipped with highly qualified personnel and the best 21st century facilities to offer culturally appropriate, holistic preventive, mental health and curative treatment services" (NAARC n.d.). The NAARC emphasizes racially appropriate care designed to see the Black body and mind thrive. The NAARC is all too familiar with the role that the US government played in the growth of the transatlantic slave trade and the fruition of chattel slavery to drive an imperialist and exploitative market.

We can further complicate our understanding of the intersection of health and the experience of slavery by thinking explicitly about the realities of

Black women who were enslaved or indentured and forcibly made to breastfeed white children. In her conversation on first food justice, legal scholar Andrea Freeman reiterates this truism: "People are made to believe that Black women are unfeeling, cold mothers who would or could not breastfeed, and who are completely distinct from the pervasive images of nurturing, breastfeeding white women who symbolize maternal best practices. The origins of this myth are in slavery, when slave owners benefited from a narrative about Black mothers that provided moral justification for wrestling them away from their own infants in order to breastfeed white babies" (Freeman 2015, 3058). The psychology of providing a priceless gift meant to bond parent and child for one's owner is unfathomable. Another legal scholar, Twila L. Perry, carefully explains the nuanced relationship between white and Black feminisms with regard to motherhood. "All too often, society portrays Black women as inadequate to the task of mothering Black children. They are portrayed as competent caretakers of white children, but only in the role of domestic worker, rather than legal mother" (1998, 104). Enslaved Black women faced an even more acute version of this emotional and physical violence, having been forcibly separated from their children and consequently required to share the gift of breastfeeding with white babies, thus investing in creating robust white children at the hands of state-sanctioned power. At this time, the ten-point plan does not mention the perinatal years, pregnancy, childbirth, or lactation, though Duggan and Abusharaf (2006), in their contribution to *The Handbook of Reparations*, remind us of the need for forms of redress that are not gender-blind. Scholarship and policy furthering reparations would be remiss to not attend to the significant contributions that Black breastfeeding has had in shaping the US and Black female consciousness with regard to mothering. Food justice and racially based liberation movements have an opportunity to expand the understanding of what food justice is and can be by partnering with existing stakeholders to center a gender-informed lens of how reparations can engage a historical breastfeeding narrative in order to move forward equitably.

Breastfeeding Is Food Justice

First food equity contributes to the food justice movement first and foremost by foregrounding gender and multiple feminisms. The medical and public health fields have created racialized mistrust over centuries of misuse of the

Black body. For this reason and more, "women report feeling most comfortable with and most likely to seek out breastfeeding guidance from health practitioners with similar experiences, understandings, and backgrounds as them, because significant research shows that there are marked disparities in how public health, medical care, and human service providers treat patients of different racial backgrounds" (Morell 2017, 491). However, over 85 percent of international board-certified lactation consultants, nurse-midwives, and nurse practitioners are white (Morell 2017, 491). Food justice has an opportunity to lead movement pedagogy in recognizing the relationships between gender and race. Kimberly Seals-Allers, nationally recognized expert in Black breastfeeding equity and author of *The Big Letdown*, leans into the food justice nexus by describing communities lacking in breastfeeding support as "First Food Deserts" (2017). The lack of racially and culturally responsive breastfeeding support in this country represents an opportunity for food justice practitioners and advocates to mobilize in the name of parents and babies. Shared language and political understanding are already a common vernacular pertaining to both food justice and first food equity. Applying a nuanced understanding of racial equity in the name of expanding the scope of food justice to include first food equity might create the conditions for the food justice movement to call in breastfeeding equity and work together.

Similar to food justice advocacy, first food equity complicates our understanding of what "local" means and for whom. In May of 2012, Dr. Vandana Shiva (n.d.), ecofeminist and food sovereignty activist, shared the following tweet: "Mothers milk is the baby's first food. It is zero mile food with no food miles, no greenhouse gas emissions." Given its intersectionalities, first food equity may also serve as a salient indicator of social progress in being able to discern the extent to which social services, urban design, and equitable first food opportunities are available for new parents and their babies. Breastfeeding lends a new understanding of localism and scope with regard to food systems. La Leche League Great Britain emphasizes aligned messaging: "Breastfeeding is one of the best ways to protect both our planet's ecosystems and our health. It helps to ensure food security, has a positive effect on maternal and child health and wellbeing, and is vastly important to our carbon footprint. There's no food more locally produced, more sustainable or more environmentally friendly than breastmilk" (quoted in Burbidge and Williams 2020). Le Leche Great Britain further explains that it's been called the most "food mile friendly" product there is. It's a naturally renewable resource and gives babies all the nutrients they need for around the first six months of life (Burbidge and Williams 2020). Ecofeminism is defined as a political and intellectual movement that combines feminism and environmentalism. It argues that the oppression of women and the degradation of the environment are interconnected and result from patriarchal and capitalist systems. Ecofeminism advocates for a shift in values toward care and cooperation to benefit both society and the environment. First food equity highlights the ways that patriarchy shapes food systems and further emphasizes the need for a feminist approach to food justice.

This critically important conversation around first food equity is gaining momentum in both peer-reviewed scholarship and community-based advocacy spaces. First food equity encompasses systemic diversification of access to breastfeeding-enhancing policies, lactation support, and community support, all designed to increase breastfeeding rates in Black, Brown, Indigenous, and low-income communities. With support from the W. K. Kellogg Foundation, HealthConnect One is working to position first food equity as part of the larger food justice movement. The First Food Equity Project aims to improve the initiation and duration of breastfeeding rates among low-income communities by expanding community-based, peer-to-peer support models that diversify both the maternal and the child health workforces rooted in these communities. "The First Food Equity project provides us with the precious opportunity to grow the community-rooted lactation support in communities across the nation where the need is the greatest," said Dr. Twylla Dillion, executive director of HealthConnect One. "For the communities we are working with, the option to work with someone with shared lived experience can be life-changing" (HealthConnect One, 2021). It will be critically important for the first food equity movement to remain cognizant and respectful of the spectrum of feeding choices available to parents and babies. Well-intentioned movements can inadvertently share messaging akin to the texts I read as a young mother—calling breastfeeding "natural" and "beautiful." While breastfeeding is the right choice for some, it is imperative that first food equity creates the opportunity for parents and babies of all backgrounds to actively choose feeding preferences that work for them.

Breastfeeding support needs to operate at scales beyond the individual family. In her article "First Food Justice: Infant Feeding Disparities and the First Food System," sociologist Erica Morell emphasizes, "It is important to take seriously breastfeeding's social, economic, and environmental impact, so too we must critically regard how society, the economy, and the environment impact breastfeeding. Indeed, if we want to reap breastfeeding's shared benefits, we must recognize and address our common role in fostering them" (2017, 491). First food equity makes several core claims. First, it positions breastfeeding as a global food issue that is deserving of investment from intersectional stakeholders. Furthermore, it continues to acknowledge the inextricable relationship between gender and race. In addition, first food equity acknowledges the unending physical, emotional, and systemic benefits that breastfeeding affords a society and posits that the work of breastfeeding should be highly valued. Furthermore, Morell aptly shares, "Breastfeeding and first foods affects [*sic*] all of us as both individuals and as a society of eaters and feeders" (2017, 491). Breastfeeding is often thought of as a personal choice enacted independently of intersectional factors. Morell continues to shed light on this common misconception: "We all affect breastfeeding. Society, the economy, and the environment constrain and enable breastfeeding success. Breastfeeding is thus an issue that shapes and is shaped by everyone. It is important that we recognize this reality for our own individual interests as well as those we as a citizenry hold in common" (2017, 491).

Planners and policymakers have taken steps toward addressing this critical need. The Milan Urban Food Policy Pact (MUFPP), an international agreement of over a hundred mayors, has begun the work of partnering food justice and first food equity by affording specific nutritional consideration to the events of the perinatal years. MUFPP's framework for action includes indicators for sustainable diets and nutrition, defined as healthy, safe, culturally appropriate, environmentally friendly, and rights-based, and pays particular attention to the needs of pregnancy-capable people as well as nutritional requirements for infants and young children (MUFPP 2018).

In economic terms, breastfeeding is an unpaid, system-level health care contribution by women and pregnancy-capable parents. Unpaid health and child care provided in the household, along with other socially reproductive activities that contribute to the physical, cognitive, and emotional development of members of a household, have a major impact on individual and public well-being, as well as on the human development potential of countries. These economic activities take place outside the market and are therefore invisible in the economic statistics and national accounts systems of most countries (Smith 2019, 480).

Breastfeeding is by no means free; it comes with significant financial opportunity costs that few government policies are equipped to mitigate. *Forbes* (Epker 2023) pursued a deep dive into the true value of unpaid, largely women's work. One estimate of the globally aggregated "true economic value of this work" is $10,900,000,000,000. In other words, if women around the world received the minimum wage for every hour of their unpaid labor, they would have contributed about $10.9 trillion to the global economy in 2020, a figure that is more than twice the size of the global tech industry for that year ($5.2 trillion). If women in the US alone earned the minimum wage for their unpaid work, they would have made about $1.5 trillion collectively in 2019 (Upwardly Global 2025). "The value of breast milk, breastfeeding and other socially reproductive labor should be seen as a serious omission in the value of home produced goods and services" (Smith 2019, 480). Family-enabling policies including the Family Medical Leave Act, Parental Universal Basic Income, subsidized and culturally appropriate doula and lactation support, family-friendly work policies, and affordable childcare each offer opportunities to afford new parents access to evidence-based care and social independence. Policy designed to mitigate the burdens of parenthood, be they physical, emotional, financial, or otherwise, serve to advance the fruition of first food equity. The principles of first food equity stand to inform family-enabling policies, thus allowing parents to actualize their desired feeding preferences with similar outcomes irrespective of racial, cultural, or economic background.

Conclusion

Breastfeeding is a food justice issue involving our first and perhaps most important food. A partnership between food justice and first food equity would acknowledge the vulnerability of infants and toddlers and afford our very youngest community members the respect they and their parents deserve in order to thrive. Legal scholar Andrea Freeman succinctly shares, "Food oppression is institutional, systemic, food related action or policy that physically debilitates a socially subordinated group" (Freeman, 2015, 3066). First food equity stands to mitigate the systemic boundaries that contribute to support for all infant feeding choices. As suggested in this chapter we have a plethora of tools with which to understand breastfeeding's role within pluriversal societies. Feminist theory, Black feminist theory, food justice, environmentalism, economics, and community-based education represent multiple approaches that we can leverage in pursuit of first food equity. We have a critical opportunity to model movement-based best practices by inviting food justice to incorporate breastfeeding justice and first food equity into an expanded and newly defined scope of what food justice actually encompasses.

References

Allen, P., and Sachs, C. (2007). "Women and Food Chains: The Gendered Politics of Food." *International Journal of Sociology of Agriculture and Food* 15(1): 1–23. https://doi.org/10.48416/ijsaf.v15i1.424.

Buckingham, S. (2015). "Ecofeminism." In *International Encyclopedia of the Social & Behavioral Sciences*, 2nd ed., 845–850. Elsevier.

Burbidge, A., and Williams, E. (2020). "Breastfeeding and the Environment." La Leche League. https://laleche.org.uk/breastmilk-and-the-environment/.

De Grieff, P. (2006). *The Handbook of Reparations: International Center for Transitional Justice*. Oxford University Press.

Duggan, Colleen, and Adila M. Abusharaf. "Reparation of Sexual Violence in Democratic Transitions: The Search for Gender Justice." In *The Handbook of Reparations*, edited by Pablo de Greiff, 623–649. Oxford University Press, 2006. https://doi.org/10.1093/0199291926.001.0001.

Epker, Eva. (2023). "Women Handle 75%+ of All Unpaid Labor. Their Health Pays the Price. *Forbes*. October 31. https://www.forbes.com/sites/evaepker/2023/10/31/women-handle-75-of-all-unpaid-labor-their-health-pays-the-price/.

Food and Agriculture Organization of the United Nations, MUFPP, and RUAF. (2018). "Milan Urban Food Policy Pact Monitoring Framework Indicator 7: Minimum Dietary Diversity for Women of Reproductive Age." https://openknowledge.fao.org/server/api/core/bitstreams/84b48e0b-4900-498b-87c9-995c49b13253/content.

Freeman, Andrea. (2015). "'First Food' Justice: Racial Disparities in Infant Feeding as Food Oppression." 83 *Fordham Law Review* 83(6): 3053–3087. https://ir.lawnet.fordham.edu/flr/vol83/iss6/8.

HealthConnect One. (2015). "Read The Perinatal Revolution Report (2014)." https://healthconnectone.org/publication/read-the-perinatal-revolution-report-2014/.

Henigsman, S. A. (n.d.). "How Many Eggs Are Women Born With? And Other Questions About Egg Supply." Healthline. Retrieved October 7, 2024. https://www.healthline.com/health/womens-health/how-many-eggs-does-a-woman-have.

Karp, H. (n.d.). "What Is the Fourth Trimester of Pregnancy?"Happiest Baby. Retrieved October 7, 2024. https://www.happiestbaby.com/blogs/baby/fourth-trimester.

Mies, M., and Shiva, V. (2014). *Ecofeminism*. Bloomsbury.

Milan Urban Food Policy Pact. (n.d.). "Sustainable Diets and Nutrition." Retrieved October 19, 2024. https://www.milanurbanfoodpolicypact.org/framework-for-action/sustainable-diets-and-nutrition/.

Morell, E. (2017). "First Food Justice: Infant Feeding Disparities and the First Food System." *Breastfeeding Medicine* 12(8): 489–492.

National African American Reparations Commission (NAARC). (n.d.). "What is NAARC's 10-Point Reparations Plan?" Retrieved October 6, 2024. https://reparationscomm.org/reparations-plan/.

Perry, Twila L. (1998). "Transracial and International Adoption: Mothers, Hierarchy, Race, and Feminist Legal Theory." *Yale Journal of Law and Feminism* 10(1): 101–164.

Porter, R. (n.d.). "10 Facts You Didn't Know About Breastfeeding." Western Missouri Medical Center. Retrieved October 7, 2024. https://wmmc.com/ten-facts-about-breastfeeding/.

Reese, A., and Cooper, D. (2021). "Making Spaces Something Like Freedom: Black Feminist Praxis in the Re/Imagining of Just Food System." *International Journal for Critical Geographies* 20(2): 450–459.

Sachs, C., and Patel-Campillo, A. (2014). "Feminist Food Justice: Crafting A New Vision." *Feminist Studies* 40(2): 396–410.

Seals-Allers, Kimberly. 2017. *The Big Letdown: How Medicine, Big Business, and Feminism Undermine Breastfeeding*. St. Martin's Press.

Shiva, V. (n.d.). "Dr. Vandana Shiva (@drvandanashiva)." X.com. Retrieved October 13, 2024. https://x.com/drvandanashiva.

Smith, J. (2019). "Counting the Cost of Not Breastfeeding Is Now Easier, but Women's Unpaid Health Care Work Remains Invisible." *Health Policy and Planning* 35: 479–481. https://doi.org/10.1093/heapol/czz064.

Upwardly Global. 2025. "About Economic Opportunity." January 28. https://www.upwardlyglobal.org/eo/about/ .

19 The Green New Deal and the Future of Food Justice

Maggie Dickinson

As both scholars and activists have noted, the Green New Deal offers an unprecedented opportunity for food justice movements to overcome some of the historic challenges to realizing a more just food system. Introduced as a congressional resolution in 2019, the Green New Deal presents a framework to transform the US economy to reach "net-zero greenhouse gas emissions through a fair and just transition for all communities and workers" (Ocasio Cortez and Markey 2019). Food and agriculture are named as systems in need of transformation in order to both reduce carbon emissions and ensure access to sufficient, healthy food for all people.

The Green New Deal framework offered a breakthrough opportunity to think big about food system change and the future of food justice. Scholars studying food justice movements and grassroots organizations consistently note the disconnect between the "radical visions" and "neoliberal constraints" of food justice efforts (Alkon and Mares 2012), including issues of scale, the realities of an austere philanthropic environment, and the internal race and class dynamics that often center white and middle-class perspectives and concerns (Reynolds and Cohen 2016). In 2019, the Green New Deal seemed to offer a vision that broke through some of the neoliberal constraints of what is possible—at least in relationship to scale and the ability to move past philanthropic funding models—to put the full force of the federal government behind food system transformation. However, many tensions remain. Building on recent scholarship on food justice and abolition (Montenegro de Wit 2021; chapter 20, this volume), I ask how competing perspectives on racism and capitalism shape political imaginaries and organizing that inform what a Green New Deal (GND) could look like.

In particular, I want to think about the ways that attending to or failing to attend to gendered racial capitalism structure policy analysis. Ashanté Reese and Joshua Sbicca, drawing on Eric Holt Gimenez, point out that "too often, food activists, especially those with greater privileges, frame the food system as broken and failing people and the environment", rather than as a system where exploitation and degradation of land and labor are central to how our food system functions (Reese and Sbicca 2022). That is to say, there is often a disconnect between people who see the need for reforms to the current food system that leave the basic structures of landholding, exploitative labor relations, and private ownership of capital established under plantation agriculture in place and those who see the need to "change everything" in order to rebuild the world, in the words of abolitionists like Ruth Wilson Gilmore (Gilmore 2023).

This split was on display at a meeting I helped pull together in New York City in 2019 in an attempt to build coalitions across the food system to generate a food justice agenda for a state- and city-level GND. I'll lay out a number of positions on display at that meeting to help pull apart the political tensions at play in conversations around food and the GND and think through how abolitionist methodologies can guide our approach to food policy work. As a white settler, I take an auto-ethnographic approach to look at some of the limitations in my own thinking and organizing and the ways I was pushed to expand my political horizons to consider what is necessary to achieve a just transformation in the food system. All the participants quoted in this piece were sent the final draft manuscript and asked permission to quote them. I offered to edit or delete any quotes they preferred not to have published. I am especially grateful to Dennis Derryck, whose contributions at the meeting and in his comments on this piece after it was written have significantly shaped much of the analysis below. His contributions were crucial in unearthing many of the unspoken assumptions at play in policy work around a GND.

By looking carefully at some of the tensions that emerged at this meeting, I illuminate how our personal shaping—as people living in a colonial, capitalist, and white supremacist world—can show up in our approach to policy work. Following adrienne marie brown's insights into the fractal nature of our organizing work—the realization that what we practice at the microscale is reproduced at all levels of the work we do (brown 2017)—I ask how our implicit and explicit understandings of the role racism plays in the production of a food system that is pushing the world toward climate crisis shape how we imagine what a GND might look like and what it can achieve. When we look at the kind of proposals that emerge around concepts like a Green

New Deal—which range from the very small to the transformative—we must ask, What is preserved? What healing is made possible by these interventions, and what healing is foreclosed? In the building of a new world—one that none of us have seen in our lifetimes—what guides our experimentations and imaginings about food justice?

Deepening Our Understanding of the Process, the Problems, and the Possibilities

Working with colleagues at the CUNY Urban Food Policy Institute, I invited scholars, activists, and practitioners to participate in a meeting to ask what kinds of food and farm policies were necessary to realize a just transition in the food system. The thirty-five participants, who included farmers, urban gardeners, scholars and antihunger, labor, and other food organization leaders from New York City and the state submitted their top three to five policy priorities for a state, city, or federal GND as the basis for the discussion prior to our meeting. We had a few keynote speakers, and then participants worked in groups (agriculture, hunger and nutrition, food waste, and purchasing and procurement) to see if we could refine a set of principles and policy priorities.

There were admittedly mistakes in how we organized: we did not offer stipends or childcare support, and we planned the meeting for a Friday in late May—a terrible time for farmers and a bad day for people who have to go to work. In her keynote address Onika Abraham, executive director of Farm School NYC and a Black food justice activist, asked how many of us in the room were being paid to be there, and nearly everyone raised their hand. Implicit in her question was a critique of an organizing model that was top-down and thinly rooted. There were a number of ways our meeting excluded people, which shaped the conversations that took place and the efficacy of these efforts. In many ways, it was reflection on these shortcomings that helped me to really understand the tensions that emerged in this gathering and the ways they were rooted in divergent understandings of the problems that we face. These are not secondary issues, but, as I hope to show, are reflective of ways of being and understandings of power that need to be critically assessed.

The participants who pointed out the limitations in our organizing were the same participants who clearly rooted their policy analysis in colonialism, slavery, and dispossession and put forward calls for reparations—particularly land reparations for Black farmers and Indigenous communities. In his book *Reconsidering Reparations*, Olufemi Táíwò points out, "If slavery and colonialism built the world and its current basic scheme of social injustice, the proper task of social justice is no smaller: it is, quite literally, to remake the world. In our era, climate justice and reparations are the same project: climate crisis arises from the same political history as racial injustice and presents a challenge of the same scale and scope. The transformations we succeed or fail to make in the face of the climate crisis will be decisive for the project of racial justice and vice versa" (Táíwò 2022). Similarly, a number of participants at our meeting were clear in articulating that slavery and colonialism built our food system and structure the forms of environmental degradation, dispossession from land, exploitative labor, and unequal experiences of hunger that are endemic to our current food system.

More than that, these same participants were deeply attentive to the relational dynamics of what must be built, nurtured, and nourished as part of a just transition. In her keynote address, Abraham noted the difference between strong sustainability and weak sustainability. She pointed out that "strong sustainability acknowledges that there are certain functions that the environment performs that cannot be duplicated. So our ozone layer can't be replaced. Our soil takes eons to produce. Things in the community that cannot be replicated—by saying Amazon can feed us. That replaces our community relationships. Nothing is wasted, nothing is disposable—especially people." She and others expressed concerns about the people left out of the conversation and the need to rethink how we do this type of policy work.

A number of participants came to the gathering with specific policies that they were advocating. My own focus was the need to end exploitative labor relations in the food system through a federal jobs guarantee that would provide life-sustaining jobs to anyone who needs or wants one, give people an alternative to exploitative conditions in the private sector, and create new possibilities to staff and scale up food justice work (Dickinson and Lohnes 2019; Paul et al. 2018). Others were focused on price parity for farmers, good food purchasing programs, including farmworkers in labor law protections, establishing community land trusts, breaking up monopolies across the food system, encouraging vegetarianism, reducing food waste, and numerous other policy proposals.

I came into this gathering, perhaps naively, imagining that we could start with policy—the what is to be done—and move on from there. My own perspective coming into this meeting was shaped by what I saw as a political opportunity. I saw a federal job guarantee as the most transformative demand in the Green New Deal framework—one that had the potential to scale up grassroots food justice efforts at an unprecedented speed and scale. Dennis Derryck, founder of Corbin Hill Food Project and long-time Black food justice activist in New York City, challenged my focus on "good jobs" in the GND framework, noting that it did nothing to address the loss of land in Black and Brown communities. He pointed out that the job guarantee framing simply would not get us where we need to go in undoing the racist landholding and land use patterns that undergird our entire food system. Pointing this out wasn't intended to create a binary between good jobs and land reform. Ultimately, a federal job guarantee was an idea he supported. His point about the limits of this policy was intended to push our (and specifically my) horizons further in terms of the root causes of injustice in the food system and how they need to be addressed.

The Dangers of Urgency

At the other end of the spectrum, a small number of participants questioned these priorities around land redistribution and a job guarantee as distractions from the urgent need to reduce carbon in the atmosphere by any means necessary. Todd Fernandez, executive director of Climate Crisis Policy and a white-presenting climate change activist, suggested that programs like a federal job guarantee and land reparations were distractions from the urgent work of reducing carbon in the atmosphere. His position (and mine to some degree) presumed that we know how and what to build/reform/transform to achieve a narrow, measurable target. He appealed to pragmatism and urgency in the face of imminent climate crisis. His disregard of the need to transform power dynamics in the labor market and in relationship to land as irrelevant to the pressing question of sequestering carbon—a well-intentioned goal shared by everyone at the meeting—reflects the "use of crisis as a way to stabilize systems in place that puts certain things beyond dispute in the imperative to act at all costs" (Liboiron 2021). His dismissiveness of certain proposals as beyond the pale or distractions to the urgent need to sequester carbon stood in contrast to Abraham's emphasis on who was left out or Derryck's attempts to find common ground while at the same time pushing others to expand their understanding of the fundamental role racism has played in producing our ecologically destructive food system.

But what happens when we attend to the urgent need to sequester carbon—for example, by incentivizing new farming methods—without attending to the relational dynamics embedded in landholding patterns? Indigenous scholars like Nick Estes, one of the thinkers behind the Red Deal, remind us that the kinds of scaling up done as part of the original New Deal—for example, the building of dams—dispossessed Indigenous communities and furthered the genocidal politics of the US (Estes 2019). Social security and labor protections—hallmark programs of the original New Deal—were designed to exclude domestic workers and agricultural workers to maintain white control over this largely Black labor force in the US South (Katznelson 2005). The exclusion of agricultural workers from labor protections in the original New Deal reverberate today in the ongoing challenges farmworkers face in collective organizing for better wages and working conditions. This sense of urgency runs the risk of pushing for policies that further entrench a deeply unequal food system in the name of saving the planet.

The urgency to prioritize and push forward "realistic" or "achievable" policy proposals runs the risk of reproducing what Thompson and Watson term "advertisements for the state," or the understanding of the state as an entity capable of solving these problems rather than as a nexus of power that must be engaged judiciously. They note that "one problem with such advertisements is that without an accurate sense of history, white people are unable to imagine a future that is fundamentally different from the present" (Thompson and Watson). The resultant flattening of history and lack of creativity and imagination limit our horizons of what is possible and desirable. Fernandez was articulating a view that we could reduce carbon without fundamentally reorganizing the basic structure of the food system as it currently exists. When we were in the room together, I felt the pull of his sense of urgency and shared it to some degree. This shared sense of urgency raises a number of questions: Why the urge to gloss over, to not know, to fail to acknowledge or see? What is it that we don't want to know? And, perhaps most importantly, what happens when we engage these questions—what becomes possible?

Finding an Abolitionist Political Horizon

I'm obviously not the first person to say this, but it bears repeating that our food system and its contributions to climate catastrophe are a set of interrelated technical and relational problems that need to be addressed simultaneously. The growing field of Black ecologies, drawing on theorists like Sylvia Wynter and Katherine McKittrick, calls our attention to how our contemporary food system is rooted in plantation agriculture and racialized violence and also how Black resistance and opposition to these forms of domination offer "assemblages that lead out of socioecological crises toward better futures" (Davis et al. 2019; Wynter and McKittrick 2015; Williams and Porter 2022; Freshour and Williams 2022; Nation 2021). This is where I think abolitionist methodologies around policy analysis are crucially important for thinking about the myriad interventions necessary to realize a just transition. When it comes to food systems, I suggest that abolitionist methodologies for policy work ask us to consider how various proposals to reduce carbon in the atmosphere shape people's relationship to land, labor, and food. Rooted in an analysis of racial capitalism, abolitionists center the harms caused by the racist enclosures of policing and prisons and the need for a renewed commons (Kaba and Ritchie 2022). Similarly, by centering the harm caused by the enclosure of land, labor, water, food, and seeds in the food system, we are able to determine a new horizon that extends beyond the need to simply reduce carbon in the atmosphere.

Abolitionists engage in concrete policy work that centers nonreformist reforms, or "those measures that reduce the power of an oppressive system while illuminating the system's inability to solve the crises it creates," from campaigns to end solitary confinement and the death penalty to developing new, nonpunitive modes of conflict resolution (Berger, Kaba, and Stein 2017). When it comes to food systems, I suggest that abolitionist methodologies for policy work ask us to consider the oppressive systems that structure the food system and how various proposals to reduce carbon in the atmosphere may or may not reduce the power of these systems. Abolitionists refuse reforms that strengthen or uphold violent systems of oppression. For police and prison abolitionists today, this means opposition to any policy that shifts more funding, resources, or legitimacy to prisons or policing. At the same time, abolition is world-making project and practice to create "the conditions that would allow for the dismantling of prisons, policing and surveillance and the creation of new institutions that actually work to keep us safe and are not fundamentally oppressive" (Kaba 2021). Abolition offers a framework for asking what is to be done that centers human well-being and divests from systems of harm and violence. What constitutes a nonreformist reform can be highly contextual and challenging to sort through. By keeping the relational dynamics of any intervention at the forefront—that is, by always asking how interventions or policies can move us away from harm and by asking how policies might move resources into the kind of world-building efforts needed to make violent systems of control obsolete—abolition offers a methodological framework for how to critically relate to the messy and imperfect world of policy work. The activists and organizers at our GND gathering who insisted on turning our attention to the long history of racist land dispossession and labor exploitation were offering a framework for assessing the various policy proposals that emerged in our conversations.

What Derryck and others were challenging us to do by invoking the long history of colonialism, enslavement, and dispossession from the land was to tell a different story—one that pushed us to reconsider what was both possible and necessary. They were pointing us toward an abolitionist horizon by insisting on a story about food and climate that took us back to colonialism and the relationships to land and racialized bodies that are the foundation of our current food system. This type of storytelling is an alternate mode of theorizing (Vasudevan et al. 2022) that can help us resituate contemporary planetary crises within longer histories and plural understandings of our relations with the earth (Sultana 2022). In doing so, these participants posited a horizon of freedom, realizable only with the redistribution of land—or even further, the abolition of property itself (Walcott 2021). They were introducing "unthinkable" or "impermissible" thoughts (Shange 2022) into the conversation.

Unthinkable or impermissible thoughts are closely connected to the ways white supremacy is embedded in policy work. Thompson and Watson note the "historical amnesia" that maintains white supremacy: "a scant or misinformed understanding of history makes it easier to pretend that the past did not happen or that it has little meaning for today, which makes it possible to deny responsibility for what has happened and for enacting new ways of being" (Thompson and Watson). My sense of urgency in this meeting was shaped by a lack of historical depth in my own analysis of what was possible. By pushing me to expand the historical scale through which we conceptualized the problem of our unequal, carbon-intensive, death-dealing food system, the systems that must be dismantled, and the many harms that must be repaired, Derryck and others were offering a new common sense about what technical, relational, and political horizons we need to work toward. In doing this, these participants offer us a way to evaluate the cracks that can be widened and stretched to build alternatives and to dismantle oppressive systems by paying attention to the new "socioecological realities" that are made possible (Davis et al. 2019).

Doing Food Policy: Finding the Breaks and Stretching the Cracks

Savannah Shange writes that "abolition is about attending to the break—stretching open the cracks in captivity's infrastructure" (Shange 2022). She offers a way to think about policy in relationship to a Green New Deal as an opportunity to both constrain the harm caused by our current food system and to stretch the cracks where something new can grow. Doing so requires us to look for the cracks and the breaks in the racialized systems built through the plantation, colonialism, and white supremacy. Guiding questions for this work include What does any particular policy make possible, and how does it move us to an abolitionist horizon? Does it make room for people to build something necessary—including new relationships with one another and to the land? Does it break, drain resources from, or confront something that is perpetuating harm and must be abolished?

The introduction of a Green New Deal framework in Congress sparked an explosion of policy proposals, but the work to transform the food system did not begin or end there. In our gathering, people were drawing on the work and insight of people and organizations who are already deeply engaged in making new worlds and dismantling existing power structures. Platforms like HEAL Food Alliance's Platform for Real Food (HEAL 2018) and Soul Fire Farm's Food Sovereignty Action Steps (Farm 2019) had already done much of the intellectual labor of identifying both the historic roots of our unequal, unsustainable food system and the policies needed to move us in a different direction. A number of the people attending our meeting were involved in these networks and this ongoing political imagining.

These platforms and the policy proposals that emerged from our work share a number of common threads. One of these is a divest/invest approach (Movement for Black Lives n.d.) to federal agricultural subsidies that would divest funding from ecologically devastating farming practices on the part of large corporate landholders and invest federal funding in agroecological farming practices that both sequester carbon and address the relational aspects of landholding. Proposals for increased federal funding to transform farm practices are paired with language that addresses the specific challenges farmers of color face—including foreclosure, a lack of support getting loans and navigating bureaucratic systems, and access to land (Minkoff-Zern 2019). These are policy approaches that identify the cracks that can be widened and the breaks in the food system that can be expanded.

Here, again, we can see the ways that the tensions captured at this meeting reverberate out into the ongoing policy efforts to decarbonize the food system. The sense of urgency Fernandez expressed shows up clearly in the policy approach his organization has taken since this meeting occurred. They have drafted the Earth Bill, which would legislate 100 percent regenerative agriculture by 2030 and which has been introduced into the US House of Representatives. It lays out a framework in which regenerative agriculture is defined exclusively through technical processes. Federal grants would pay farmers and ranchers to implement regenerative farming methods like silvopasture and cover crops. Though this bill draws on the policy platforms that informed our meeting, as well as others that have since advanced and expanded a divest/invest approach, there is no mention of or mechanism for addressing the deeply racist landholding patterns in the US, where "the top five landowners, all white, own more acreage than all of black America combined, and 98 percent of agricultural land acreage is white-owned" (Van Sant 2019). The Earth Bill offers a framework for sequestering carbon without acknowledging the racist underpinnings of landholding patterns and corporate consolidation that produced our unsustainable agricultural system in the first place.

If realized, this bill would become a boon to large landholders in ways that also entrench current power dynamics in the food system. Abolitionist methodologies ask us to consider whether or not policies preserve, expand, or legitimize systems of harm. Given the role of landholding and consolidation in carbon-intensive agricultural systems, offering subsidies and technical support to farmers and ranchers without breaking up monopoly landholding will bolster the power and legitimacy of large landholders, furthering W. E. B. Dubois's observation that "whiteness is the ownership of the earth forever and ever, Amen!" (Dubois 1920). This is not to oppose the goal of carbon sequestration—but to acknowledge the limits and short-sightedness of an approach that does not center racism, land consolidation, and the longer history of power dynamics that produced our current food system. Rather than stretching the cracks, this approach papers them over and strengthens existing relationships to land through subsidizing wealthy white land owners' investments in and control over land and food production.

Looking at the ways the small interactions I document from our gathering reverberate into the kinds of policy that get produced helps us see how easily racism, inequality, and labor exploitation are pushed aside when they are seen as distractions rather than constitutive of the problems we face. My hope is that by thinking out loud about the ways that we sometimes come up short and by taking seriously the intellectual labor of activists and advocates who are painstakingly engaged in building the world we need, we—and here I am thinking particularly of white activists and scholars—can learn to listen deeply and reflect on the kinds of policy work we engage in. As Bell et al. in chapter 13 of this volume show, it is important to see that good intentions are no substitute for grounded historical analysis that accounts for the ways our food system is designed to extract resources and labor from the earth and from racialized bodies. A Green New Deal offers an opening for food system transformation, but to make the most of it we need a clear understanding of the relational dynamics at the heart of our unequal, unsustainable food system—not as secondary issues, but as the very heart of the matter.

References

Alkon, Alison Hope, and Teresa Marie Mares. 2012. "Food Sovreignty in US Food Movements: Radical Visions and Neoliberal Constraints." *Agriculture and Human Values* 29: 347–359.

Berger, Dan, Mariame Kaba, and David Stein. 2017. "What Abolitionists Do." *Jacobin*, August 24, 2017. https://jacobin.com/2017/08/prison-abolition-reform-mass-incarceration.

brown, adrienne marie. 2017. *Emergent Strategy: Shaping Change, Changing Worlds*: AK Press.

Davis, Janae, Alex Moulton, Levi Van Sant, and Brian Williams. 2019. "Anthropocene, Capitalocene, . . . Plantationocene?: A Manifesto for Ecological Justice in an Age of Global Crises." *Geography Compass* 13 (5). https://compass.onlinelibrary.wiley.com/doi/full/10.1111/gec3.12438.

Dickinson, Maggie, and Joshua Lohnes. 2019. "Where Do Food Banks Fit in to the Fight for a Green New Deal?" Civil Eats. May 16. https://civileats.com/2019/05/16/where-do-food-banks-fit-in-to-the-fight-for-a-green-new-deal.

Dubois, W. E. B. 1920. *Darkwater: Voices from Within the Veil*. Harcourt Brace.

Estes, Nick. 2019. "A Red Deal." *Jacobin*. https://jacobin.com/2019/08/red-deal-green-new-deal-ecosocialism-decolonization-indigenous-resistance-environment.

Freshour, Carrie, and Brian Williams. 2022. "Toward 'Total Freedom': Black Ecologies of Land, Labor, and Livlihoods in the Mississippi Delta." *Annals of the American Association of Geographers* 113 (7): 1563–1572.

Gilmore, Ruth Wilson. 2023. *Change Everything: Racial Capitalism and the Case for Abolition*. Abolitionist Papers Series. Haymarket Books.

HEAL. 2018. "HEAL Platform for Real Food." HEAL Food Alliance. https://healfoodalliance.org/platformforrealfood/.

Kaba, Mariame. 2021. *We Do This 'Til We Free Us: Abolitionist Organizing and Transformative Justice*. Haymarket Books.

Kaba, Mariame, and Andrea Ritchie. 2022. *No More Police: A Case for Abolition*. New Press.

Katznelson, Ira. 2005. *When Affirmative Action Was White: The Untold History of Racial Inequality in Twentieth Century America*. W.W. Norton.

Liboiron, Max. 2021. *Pollution Is Colonialism*. Duke University Press.

Minkoff-Zern, Laura-Anne. 2019. *The New American Farmer*. MIT Press.

Montenegro de Wit, Maywa. 2021. *Abolitionist Agroecology, Food Sovereignty and Pandemic Prevention*. Daraja Press.

Movement for Black Lives. n.d. Movement for Black Lives Policy Platform. https://m4bl.org/policy-platforms/.

Ocasio Cortez, Alexandria, and Edward Markey. 2019. "Recognizing the Duty of the Federal Government to Create a Green New Deal." In House Resolution 109. Washington DC. https://www.congress.gov/bill/116th-congress/house-resolution/109.

Paul, Mark, William Darity, Darrick Hamilton, and Khaing Zaw. 2018. "A Path to Ending Poverty by Way of Ending Unemployment: A Federal Job Guarantee." *Russell Sage Foundation Journal of the Social Sciences* 4 (3): 44–63.

The Red Nation. 2021. *The Red Deal: Indigenous Action to Save Our Earth*. Common Notions.

Reese, Ashanté, and Joshua Sbicca. 2022. "Food and Carcerality: From Confinement to Abolition." *Food and Foodways* 30 (1–2): 1–15.

Reynolds, Kristin, and Nevin Cohen. 2016. *Beyond the Kale: Urban Agriculture and Social Justice Activism in New York City*. University of Georgia Press.

Shange, Savannah. 2022. "Abolition in the Clutch: Shifting Through the Gears with Anthropology." *Feminist Anthropology* 3 (2): 187–197.

Soul Fire Farm. 2019. "Food Sovereignty Action Steps." https://docs.google.com/document/d/1dt0hicyhGdJSKlC3qyE1AbG9fdDrONjUh_M_bE0KMGs/edit?tab=t.0.

Sultana, Farhana. 2022. "The Unbearable Heaviness of Climate Coloniality." *Political Geography* 99: 102638. https://www.sciencedirect.com/science/article/abs/pii/S0962629822000 52X.

Táíwò, Olúfémi O. 2022. *Reconsidering Reparations*. Oxford University Press.

Thompson, Becky, and Veronica Watson. 2018. "Theorizing White Racial Trauma and Its Remedies." In *The Construction of Whiteness*, edited by Stephen Middleton, D. Roediger, and Donald Shaffer. University of Mississippi Press.

Van Sant, Levi. 2019. "Land Reform and the Green New Deal." Dissent Fall. https://www.dissentmagazine.org/article/land-reform-and-the-green-new-deal/.

Vasudevan, Pavithra, Margaret Marietta Ramirez, Yolanda Gonzalez Mendoza, and Michelle Daigle. 2022. "Storytelling Earth and Body." *Annals of the American Association of Geographers* 113 (7): 1728–1744.

Walcott, Rinaldo. 2021. *On Property: Policing, Prisons, and the Call for Abolition*. AK Press.

Williams, Brian, and Jayson Maurice Porter. 2022. "Cotton, Whiteness, and Other Poisons." *Environmental Humanities* 14 (3):499–521.

Wynter, Sylvia, and Katherine McKittrick. 2015. "Unparalleled Catastrophe for our Species? Or, to Give Humaness a Different Future: Conversations." In *Sylvia Wynter: On Being Human as Praxis*, edited by Katherine McKittrick. Duke University Press.

20 From Carceral Food Systems to Abolitionist Food Justice

Joshua Sbicca, Carrie Freshour, Kanav Kathuria, and Sara Black

Introduction

How are the struggles for prison abolition and food justice intertwined? For us as scholars and activists, organizing focused on food systems within and beyond carceral spaces provides opportunities to understand how communities facing police occupation, mass incarceration, food apartheid, and organized abandonment can build life-affirming places beyond racial capitalism's "death-dealing displacements" (Gilmore 2002).[1] Using the lenses of carcerality and abolition, this collaborative chapter explores some of these intersections, focused specifically on land, labor, and food as vehicles for creating "freedom as a place" (Gilmore 2017; Heynen and Ybarra 2021; see chapters 4, 5, 8, 12, and 15). This starting point helps us center place-making, in particular how people organize land, resources, and social capacity against the historical processes of unfreedom that produce and reproduce hierarchy, dispossession, and exclusion.

Carcerality is a core logic underlying racial capitalist and settler colonial development, whose agents capture, dispossess, displace, and confine dehumanized groups to clear land and organize labor for extraction (Wang 2018; Estes 2019). To enforce violent divisions of "human" and racialized others, carcerality requires "place annihilation" (McKittrick 2011), the never-complete ideological and material destruction of poor, Black, and Indigenous capacities for nourishing life on land. The resulting formation of plantations, reservations, ghettos, and prisons creates territories of "human sacrifice" (see chapter 4). The history of food production and eating has been central in this regard, which in the form of capitalist food systems is part of an extractive carceral universe that destroys connections to place, consolidates power, and produces profit.

In the face of carceral conditions, there have long been abolitionist notions of freedom as a place through communal practices of evasion, fugitivity, defiance, and world-building (Hames-Garcia 2004; Rodríguez 2006; Davis et al. 2022; Winston 2021). Abolition is the insistence that it is possible to organize society without dehumanization and displacement (Kaba 2021). Abolition is therefore a tradition of generating resilient life systems that refuse racial capitalism's confining hierarchies of difference (Maynard and Simpson 2022). Our aim here is to think with the need not only for dismantling carceral logics and institutions but also for liberatory practices focused on reorganizing power relationships into life-affirming formations, with food politics and abolitionist organizing at the center.

In this chapter, we explore the contours of carcerality, abolition, and food justice. First, what can carcerality reveal about food system harms that deprive people of freedom through the exploitation and dispossession of land and labor? Not only do punitive forms of social control and dehumanization produce social divisions for racial capitalist profit and territorial control, but they move through the food system as a key reproductive social force. Second, how can abolition as a world-building practice link various sites of food and non-food struggle? Abolition aims to dismantle carceral institutions and structures of harm without reproducing carceral logics. When applied to food systems, abolition requires reimaging food justice in both its scope and its practice.

Carceral Food Systems

Food is a site for the management and exploitation of incarcerated people. Carcerality is also a central organizing logic of the food system (Reese and Sbicca 2022). From the creation of hierarchies of deservingness around who has food and what kind to the forced ideological disciplining that redemption is found primarily through work, food and carcerality are deeply entangled. In the following, we identify some of these connections, with attention given to consumption, land, and labor.

If incarceration is a method of social control and dehumanization, then carceral food is a biological necessity through which to manage people and extract capital (Godderis 2006; Smoyer and Lopes 2017). When free, our guts tell us when to eat. Our social bonds provide a livable context for enjoyment. Punitive confinement ruptures these corporeal relationships to remake individuals based on the whims of the correctional officer or guard, and in cases of privatized food, such as canteens, coerced consumers of corporately produced goods. Structured mealtimes, limited choice, forced scarcity, rotten food, and more produce a disciplined carceral space (Soble, Stroud, and Weinstein 2020). Such realities reflect the idea that the incarcerated don't deserve better. Their livingness is suspended in limbo, including the biological and cultural necessity for adequate and healthy food.

Intimately related to eating are the compelled hands that feed the fellow imprisoned. Prison kitchens are the engines that provide the bare minimum calories to reproduce carceral institutions. The long-standing obsession with idleness, itself code for individuals that don't live up to capitalist and white supremacist expectations of "productive" personhood, justifies such exploitation. Unless an incarcerated person is lucky enough to get into a limited number of educational programs or experience a physical ailment, then they will serve as interchangeable units of labor. For those forced to work in kitchens, this entails preparing food bought from corporations like Aramark and Trinity Services Group, sometimes labeled "not for human consumption," and often unfit to eat (Hatch 2019).

Land and agriculture are also sites for the carceral state to respond to the fallout of racial capitalism's production of ethnoracial and class hierarchies. In rural and formerly agriculturally dependent communities where farms have been boxed out of a corporately controlled global food system, the carceral state offers jobs and a permanent supply of people to fill cages from neighborhoods in the world's metropolises facing their own organized abandonment. The exploitation of people through incarceration requires putting surpluses like land to work and then figuring out how to turn a profit (Gilmore 2007). Surpluses of labor are furthermore slotted into penal agricultural labor. Incarcerated farm workers—there are around 660 prisons with agriculture of some kind—essentially subsidize their incarceration by growing food to feed the penal system, working as lessees for agribusiness operations and producing products for prison industries for sale on the open market (Chennault and Sbicca 2023).

Institutions of confinement are not the sole means to manage poor and abandoned communities. For instance, food apartheid is a racist regime of segregation and relegation to poor health based on the structural denial of food, a violent mode of alimentary management (see chapter 6). There wouldn't be communities living under food apartheid were it not for carceral forces like housing discrimination, blockbusting, slum clearance, and other processes of forced displacement. Policing manages who can and cannot be in certain places. Criminalization, broken windows policing, redlining, and gentrification furthermore control and exclude people from racial capitalist profits. Instead of addressing poverty, and its related dietary and social fallout, the carceral state spends money to push the criminalized out of place. Displacement softens up neighborhoods to receive more upscale palates and hipster eateries that transform foodscapes for the well-to-do. Food apartheid, like the criminal punishment system itself, separates people by race and class and enforces ideas about who is deserving of nourishment, ease, and belonging (Garth and Reese 2020).

Paying attention to the connections between food and carcerality also points to the ways that immigrants are exploited around the world. Borders are everywhere militarized. They are the physical representation of who is allowed in and out. The food system is entangled in this militarization and separation insofar as borders set the parameters for who ends up as food chain workers. Much of the food chain requires precarious immigrant labor but resents immigrants. As a result of such stigmatization, citizens benefit from cheaper food. After all, if someone is only deserving of food work that is considered menial, then they don't deserve much pay for that work. To secure cheap food requires policing immigrant food chain workers in fields, factories, and dish pits with the threat of deportation. Weak labor laws furthermore ensure the carceral power of food retail and farm bosses. Immigrant food chain workers must act in line with employer interests or face the prospect of being sent back with a criminal record (see chapter 17).

In even more extreme circumstances, there is the complete curtailment of immigrant food chain workers' freedom. In places like Florida, tomato pickers from Mexico and Central America have been forced to stay in the fields and inhumane housing conditions or face violent repercussions. Immigrant farmworkers have also been funneled through human smuggling and labor trafficking operations in near-enslavement conditions where they had to pay to get into the US and then, once working, were paid little to nothing and compelled into farming operations without consent (see chapter 18). In Thailand, the seafood sector has long been plagued by fishing operators that enslave immigrants from places like Cambodia, Indonesia, Laos, and Myanmar. Exploiting human trafficking networks and economic desperation, there are countless cases of immigrant workers being forced to work around the clock in dangerous conditions, facing possible death, torture, or other physical violence, for the global seafood industry.

Taken together, these situations illustrate deep entanglements between carcerality and land, labor, and food. We ignore the root animating forces of dispossession, displacement, and confinement at the further peril of dehumanized "others" and with the consequence that we reproduce carcerality in the food system. By articulating these connections, we ultimately aim to address what needs to be abolished and how to recognize and embark on world-building practices that stitch our movements and communities together.

Abolitionist Food Justice

In recent years, the popularization of an abolitionist framework has been accompanied by the risk of deradicalization, of stripping abolition "from a grassroots, working-class, Black base or people of color base that sought autonomy from the state," as Joy James counsels (Brown University, 2019). The work to counter co-optation and center the visionary potential of abolition therefore requires commitment to real legacies of struggle and traditions of practice forged by liberation struggles (James 2005; Berger 2014), including ongoing movements for freedom founded in mutual aid, cooperative economics, and food provisioning (see chapters 1, 5, 10, 12, 13, and 15).

Prison abolitionists have long linked their work to the traditions of organizing that enslaved people utilized to resist and overthrow plantation regimes and create places that enabled their collective nourishment on the land (Davis 2005; DuBois 1992 [1935]). In the US South, these freedom projects included many Black towns and unincorporated areas as sites of vibrant life and world-building, attempts to shield Black places from white supremacist legal and political violence. Yet the end of legal slavery did not achieve full abolition. The alignment of planters and white workers in the South as well as white industrialists in the North ultimately entrenched racial capitalism and foreclosed the creation of new social institutions for human flourishing, most importantly, universal access to full employment, education, and land. These conditions overlaid settler colonial practices of dispossession that carved up formerly Indigenous territories for white agricultural and imperial interests. Settler colonial practices undermined Indigenous legal and cultural relations and produced unfreedom in highly uneven ways that complicated the abolition of slavery.

To abolish carceral structures means to build life-affirming social and material worlds, which do not require dispossession, displacement, and confinement for their coherence (Kaba and Ritchie 2022). In this spirit, we highlight struggles employing food and land as vehicles for resistance and world-building—looking beyond political pragmatism to movements that contest carcerality in myriad ways. The examples we include do not constitute a definitive or comprehensive list; rather, they reflect movements we have been involved in, inspired by, or in solidarity with. We draw on traditions of scholar-activism rooted in collective struggle, particularly abolition geographies, of which food justice must be one such endeavor, that focus on transformations of freedom into places that are emergent from life in the wake of dehumanization (Black 2022). In the words of Cedric Robinson, "We are not possible without the encouragement, the urgency, and the requirement that we be here by those who are in fact being trampled on, being imprisoned" (Osuna 2017).

Hunger Strikes and Work Stoppages in the Carceral Belly of the Beast

In the absence of effective tools of opposition toward state violence, hunger strikes and work stoppages comprise an integral form of autonomous struggle for people held in captivity (Chase 2020; Shah 2022). The refusal of food as protest spans continents and centuries. From the mass strike at the Attica Correctional Facility following the murder of George Jackson in 1971, to death fasts of Turkish political prisoners against the introduction of high-security prisons in the early 2000s, to ongoing strikes by Palestinian political prisoners protesting their settler-colonial captor's practice of administrative detention without trial or charges, hunger strikes are inherently political. Strikers use the limited means at their disposal—their very bodies—to demand action against dispossession, confinement, and unfreedom. As Palestinian political prisoner Marwan Barghouti wrote in 2017, "Hunger striking is the most peaceful form of resistance available. It inflicts pain solely on those who participate and on their loved ones, in the hopes that their empty stomachs and their sacrifice will help the message resonate beyond the confines of their dark cells" (Barghouti 2017). Hunger strikes begin at the scale of the body, reorganizing social relations beyond the dividing power of the prison to connect communities who seek to make freedom in new ways.

Given their dependence on incarcerated labor, work stoppages also reveal how prisoners assert an embodied sense of place on their own terms. In September 2016, the prisoner-led Free Alabama Movement led the largest prison strike in the history of the US—spanning twenty-four states with the participation of over twenty-four thousand incarcerated people. Tracing connections between agrarian-based slavery, racism, capitalism, prison labor, and the prison-industrial complex, the Free Alabama Movement turned to strategies developed by enslaved people centuries prior: "'Let the Crops Rot in the Field' is a proven strategy that was passed down to us from our Ancestors from the slave plantations that was used to disrupt the economics of the field. . . . When the slave master had invested all that he owned into his next crop (prison factories), the slaves would wait until just before the harvest and rebel against the slave system by 'going on strike' and causing the crops to rot in the field" (Free Alabama Movement 2015). The prison as field becomes a site for resistance, where self-possessed incarcerated workers can enact their own refusals and imagine worlds otherwise (Chase 2020).

In contradistinction to resistance through refusal lies resistance through care: the everyday practice of sharing food in confinement to support folks who do not have enough to eat or who have differing dietary or physiological needs or simply to create or strengthen social bonds (Ugelvik 2011; Gibson-Light 2018). Institutional meals induce hunger. Many individuals are unable to afford commissary supplies to replace or supplement what prisons prepare: not all are able to secure "jobs" on the inside, have loved ones who are able to send them funds, or save income from meager or nonexistent wages to afford the commissary's exorbitant prices. This drives the creation of mutual aid and solidarity networks to help navigate prisoners' lack of access to nourishing food. As Antoin Quarles, an organizer and formerly incarcerated comrade in Baltimore puts it, "Dudes come in here without any money, without any support from the outside. It's hard to get a job in here. So, we try to help people that don't have anything . . . whatever we have we share, donate it, make sure they have their hygiene and the food they need. Cuz the prison ain't going to do it so we have to do it ourselves."[2]

Divestment from the Prison Food and Agriculture Industrial Complex

In 2018, a student-led campaign at the University of Florida began to push the Institute of Food and Agricultural Sciences (IFAS) at the university to end their reliance on prisoners to work at state agricultural experiment stations. Out of a broader divestment organization, Divest UF, emerged the Coalition Against Prison Slavery (CAPS). Given the agrarian-based oppression of plantation-style chattel slavery in the South, and the popular perception that incarceration is a new form of enslavement, the choice to run a campaign against prison slavery was deliberate.

For decades, IFAS, a federal-state-county partnership, benefited from the free labor of prisoners. Around the state, prisoners were building fences, fixing irrigation, sowing seeds, eliminating weeds, harvesting crops, maintaining research facilities, and more. For agricultural scientists, this labor was considered essential, as it saved IFAS $1.69 million a year. Once the university cut its contracts with the Florida Department of Corrections (FDC), expenditures changed from $2 an hour for prisoner labor to $15 an hour plus benefits to hire free labor.

Building critical connections is central to abolitionist world-building (Berger 2014; Davis et al. 2022; Maynard and Simpson 2022). Students learned effective strategy and framing from Gainesville International Workers Organizing Committee (IWOC), which became Florida Prisoner Solidarity. In 2018, Gainesville IWOC was organizing with prisoners to stage strikes as part of a larger national wave of prison actions. As part of this work, activists held a ten-day encampment outside FDC's Gainesville Work Camp to protest prison slavery. Out of these actions emerged new abolitionist networks of students and local residents committed to disentangling public entities from the prison industrial complex. By emailing city and county commissioners, writing op-eds, demanding public meetings, hosting teach-ins, and then mobilizing people to attend public meetings, activists convinced the Gainesville and Alachua County Commissions to end their labor contracts with FDC.

While the city and county made changes quickly, it would be several more years before UF cut their contracts with FDC. Internal opposition to more expensive agricultural research was fierce. But students through CAPS managed to generate intense pressure on administrators. Through banner drops at graduation ceremonies, tabling events, meme warfare, phone zaps, direct confrontations with UF's president, direct action at board of trustee meetings,

Figure 20.1
Protestors oppose the Florida Department of Corrections' administrative management unit proposal that would increase repression of organizing across prison walls. Photo by Jordan Martinez-Mazurek.

and more, students consistently spotlighted the university's contributions to prison slavery. Once the second wave of the Black Lives Matter movement emerged in 2020, the university was looking for an easy win. After years of organizing, UF capitulated. This revealed the power of divestment campaigns, including those related to carceral food systems, as part of prison industrial complex abolition, namely in the form of diminishing its reach and power.

Food Sovereignty and Cooperative Economics

The South in general and the Black Belt in particular remain deeply significant for food-centered abolition (see chapter 15). Extending from the abolition of slavery, freed people created Black towns, settlements, and colonies separate from white established plantation power (Roberts 2023). Furthermore, the Mississippi Delta became a pivotal site of rural Civil Rights and Black Power organizing, uniquely centered around food sovereignty (Smith 2020). In 1967, to address malnutrition, joblessness, and white supremacy, John Hatch, L. C. Dorsey, and others—most coming from sharecropping and tenant farming families who saw strength in their farming knowledge and skills—established the North Bolivar County Farm Cooperative (NBCFC) on Black-owned land in the northeastern region of the Mississippi Delta (White 2017). NBCFC's central tenets of self-determination and self-sufficiency continue across a multitude of Southern Black radical and cooperative food movements.

Carrying this tradition forward to the present, Gangstas to Growers (G2G) addresses multiple carceral crises of recidivism, gentrification, and unemployment through worker-owned coöperatives for Black working-class residents of Atlanta's West Side. For founder Abiodun Henderson, this work is simultaneously about land and education, redressing the forms of violent labor systems endured through chattel slavery for people severed from land and ways of being on that land. G2G brings agricultural knowledge to formerly incarcerated youth and provides livable wages to promote self-sufficiency, healing, and a black-run food system (The Come Up Project n.d.). G2G formed Sweet Sol, a cooperative where youth grow and harvest ingredients like onions, garlic, and turmeric for their signature hot sauce. The organization furthermore embeds such work within a network of local Black-owned farms and cooperatives like Southwest Georgia Project, Southwest Atlanta Growers Cooperative, and the West Georgia Farmers Cooperative (Just Food 2021).

Cooperative freedom-dreaming and place-making can emerge from inside carceral institutions as well. LJ, a farmer and abolitionist, organizes for food justice while incarcerated in Washington State. His first experience with farming and agriculture began in his youth after being sent to a group home in rural Washington. There, he cared for animals, initially as a form of punishment, but this grew into a pivotal experience where LJ learned about farming and growing food. This carceral exposure to farming in and of itself was not enough. He had to transform his mindset through ongoing political education around Black cooperative and agrarian traditions: "The part that messed with me the most, is that the only depictions of my ancestors working on farms that I ever seen in my life . . . were all gruesome images of slave plantations in settings that look identical to the setting where I had to build an alternative food system for prisoners. It took me many years to overcome internalized racism and oppression, to the point where I could feel solid with this work."[3] LJ remains incarcerated, serving a sentence from his youth. He is working with inside-outside networks to build a farm cooperative with others in Mississippi upon release.[4]

Self-Determination Struggles in the Face of Organized Abandonment

In the face of organized abandonment, urban agriculture has become a means for communities deprived of access to nourishing food and land to build power and self-reliance (see chapters 4 and 8). Baltimore—the birthplace of racialized housing segregation policies that decades later became the foundation for redlining—is a hotbed for such forms of resistance. Just twenty-five neighborhoods in the city form the backbone of Maryland's entire carceral project—the same neighborhoods where grocery stores are scarce and rates of chronic illnesses remain high.

Yet traditions of self-determination are strong. In the majority-Black neighborhood of Cherry Hill, an urban garden stewarded by the pan-African power institution Black Yield Institute (BYI) was the community's only source of fresh, affordable, and culturally relevant produce for years. BYI explicitly works to use land and food as part of a larger struggle. As Eric Jackson, BYI's servant-director, puts it, "Black land and food sovereignty . . . is the idea that Black people control the land and resources, and that we are stewards of and in our communities. . . . It's about preparing Black and brown people to be able to shoulder the work of providing food for ourselves and controlling what comes into our community" (Ruge 2020).

BYI actualizes its vision of Black liberation in multiple ways. For over a decade, the Cherry Hill Urban Community Garden—a 1.25-acre plot of land situated on a formerly vacant lot—provided thousands of pounds of produce for residents by restoring relationships to ancestral food and growing practices severed by food apartheid. BYI's work also extends to movement support in the form of political education programming. The organization's Black Food Research & Knowledge Creation initiative utilizes community-based participatory action research methods to produce people-centered knowledge, interventions, and policy proposals. Their political education arm, the Sankara-Hamer Academy—named after freedom fighters Thomas Sankara and Fannie Lou Hamer—offers workshops, gatherings, courses, and leadership development to raise consciousness around Afrocentric forms of resistance and world-building. And the upcoming Cherry Hill Food Co-op, first developed by BYI, aims to build communal infrastructure by serving as the neighborhood's first community-owned grocery store.

In 2021, the Housing Authority of Baltimore City abruptly served BYI and the Cherry Hill Urban Community Garden with an immediate eviction notice. Yet another example of racialized dispossession, the city claimed that the land was needed to build affordable housing—without providing any plans for its construction. After months of organizing and city-wide support, BYI was able to secure a six-month reprieve to finish the growing season. With the farm now located on a property in a neighborhood a handful of miles away, BYI's work to achieve Black food sovereignty in Baltimore continues.

Conclusions

Abolitionist efforts to expose the harms of eating in confinement, connections between prison food systems and food apartheid, and more continue to be appropriated by reformist organizations and reshaped into piecemeal policies. Absent a food politics rooted in abolitionist food justice that aims to dismantle oppressive structures, advocating for healthier meals in prisons, stronger carceral nutritional policies, and more food access points in marginalized communities, for example, remains a salve. "Kinder" and "gentler" prisons or criminalized food chain workers with a few more dollars won't upend the violence of incarceration and corporately controlled industrialized food chains. It remains crucial to not reify carceral food logics into individual solutions that ignore the structural harms of racial capitalism and the carceral state. Working toward food justice demands the formulation of diverse practices with an eye toward transforming systems.

Abolition is as much a process as an outcome, a constellation of activities rather than a singular event, strategy, or accomplishment. For us to organize food systems and food relations across differences, in the wake of harm, without reproducing social hierarchies, we need robust practices for communication, strategizing, conflict engagement, harm repair, and weathering changes over time. Drawing on the organizer Mariame Kaba (2021), abolitionist food justice will require, "a million different little experiments" that refuse distinctions between "deserving" and "undeserving" people when making decisions about the resources needed to build new worlds (Gilmore 2017). There is always an opportunity to break carceral food linkages, whether inside carceral institutions or in the food system more broadly. But how we support such endeavors reflects where we find ourselves in the historical trajectory of struggle. A politics of secession from racial capitalist food systems without an abolitionist analysis that challenges carcerality in all its places and forms will be incapable of achieving food justice.

Notes

1. Our chapter's methods entail a critical engagement with carcerality and food connected to our activism and scholarship both collaboratively and independently. We are each positioned differently. Joshua Sbicca, an associate professor, has engaged in solidarity work with formerly incarcerated people through planting justice and scholar-activist work in the Prison Agriculture Lab. Carrie Freshour, currently an assistant professor, has organized with and learned from inside/outside abolitionists across Washington, Mississippi, and Georgia. Kanav Kathuria is an educator in carceral spaces and a founding member of the Maryland Food & Prison Abolition Project, a community-based organization rooted in Baltimore that interrogates carceral food systems through an abolitionist lens. Sara Black is a recent doctoral graduate who has organized for a decade with the Freedom Food Alliance, a network of abolitionist farmers in New York.

2. Antoin Quarles, personal correspondence with Kanav Kathuria, May 2020.

3. LJ, e-message to Carrie Freshour, June 2021.

4. If you want to learn more or correspond with LJ directly, please send an email to cfreshour3@gmail.com.

References

Barghouti, Marwan. 2017. "Why we are on hunger strike in Israel's prisons." *New York Times*, April 16, 2017.

Berger, Dan. 2014. *Struggle Within: Prisons, Political Prisoners, and Mass Movements in the United States*. Oakland, CA: PM Press.

Black, Sara Thomas. 2022. "Abolitionist food justice: Theories of change rooted in place- and life-making." *Food and Foodways* 30(1–2): 123–141.

Brown University. 2019. "Joy James: The Architects of Abolitionism." https://www.youtube.com/watch?v=z9rvRsWKDx0.

Chase, Robert T. 2020. *We Are Not Slaves: State Violence, Coerced Labor, and Prisoners' Rights in Postwar America*. Chapell Hill: University of North Carolina Press.

Chennault, Carrie, and Joshua Sbicca. 2023. "Prison agriculture in the United States: Racial capitalism and the disciplinary matrix of exploitation and rehabilitation." *Agriculture and Human Values* 40(1): 175–191.

The Come Up Project. (n.d.). "Who we are." Gangstas to Growers (G2G). Accessed June 1, 2023. http://www.gangstastogrowers.com/.

Davis, Angela Y. 2005. *Abolition Democracy: Beyond Empire, Prisons, and Torture*. New York: Seven Stories Press.

Davis, Angela Y., Gina Dent, Erica R. Meiners, and Beth E. Richie. 2022. *Abolition. Feminism. Now*. Chicago: Haymarket Books.

DuBois, W. E. B. 1992 [1935]. *Black Reconstruction in America: 1860–1880*. New York: Free Press.

Estes, Nick. 2019. *Our History Is the Future: Standing Rock Versus the Dakota Access Pipeline, and the Long Tradition of Indigenous Resistance*. New York: Verso Books.

Free Alabama Movement. 2015. "Let the Crops Rot in the Field: A Call for New Strategy in the National Movement against Mass Incarceration and Prison Slavery." https://freealabamamovement.wordpress.com/2015/02/26/let-the-crops-rot-in-the-fields/.

Garth, Hanna, and Ashanté M. Reese, eds. 2020. *Black Food Matters: Racial Justice in the Wake of Food Justice*. Minneapolis: University of Minnesota Press.

Gibson-Light, Michael. 2018. "Ramen politics: Informal money and logics of resistance in the contemporary American prison." *Qualitative Sociology* 41(2): 199–220.

Gilmore, Ruth Wilson. 2007. *Golden Gulag: Prisons, Surplus, Crisis, and Opposition in Globalizing California*. Berkeley: University of California Press.

Gilmore, Ruth Wilson. 2017. "Abolition geography and the problem of innocence." In *Futures of Black Radicalism*, edited by Gaye Theresa Johnson and Alex Lubin, 225–240. New York: Verso.

Godderis, Rebecca. 2006. "Dining in: The symbolic power of food in prison." *Howard Journal of Criminal Justice* 45(3): 255–267.

Hames-Garcia, Michael Roy. 2004. *Fugitive Thought: Prison Movements, Race, and the Meaning of Justice*. Minneapolis: University of Minnesota Press.

Hatch, Anthony Ryan. 2019. "Billions served: Prison food regimes, nutritional punishment, and gastronomical resistance." In *Captivating Technology: Race, Carceral Technoscience, and Liberatory Imagination in Everyday Life*, edited by Ruha Benjamin, 67–84. Durham, NC: Duke University Press.

Heynen, Nik, and Megan Ybarra. 2021. "On abolition ecologies and making 'freedom as a place.'" *Antipode* 53(1): 21–35.

James, Joy, ed. 2005. *The New Abolitionists: (Neo) Slave Narratives and Contemporary Prison Writings*. Albany: State University of New York Press.

Kaba, Mariame. 2021. *We Do This 'Til We Free Us: Abolitionist Organizing and Transforming Justice*. Chicago: Haymarket Books.

Kaba, Mariame, and Andrea J. Ritchie. 2022. *No More Police: A Case for Abolition*. New York: New Press.

Maynard, Robyn, and Leanne Betasamosake Simpson. 2022. *Rehearsals for Living*. Chicago: Haymarket Books.

McKittrick, Katherine. 2011. "On plantations, prisons, and a Black sense of place." *Social & Cultural Geography* 12(8): 947–963.

Osuna, Steven. 2017. "Class Suicide: The Black radical tradition, radical scholarship, and the neoliberal turn." In *Futures of Black Radicalism* edited by Gaye Theresa Johnson and Alex Lubin. New York: Verso Books.

Reese, Ashanté M., and Joshua Sbicca. 2022. "Food and carcerality: From confinement to abolition." *Food and Foodways* 30(1–2): 1–15.

Roberts, Andrea. 2023. "What Are Freedom Colonies?" The Texas Freedom Colonies Project. https://www.thetexasfreedomcoloniesproject.com/.

Rodríguez, Dylan. 2006. *Forced Passages: Imprisoned Radical Intellectuals and the US Prison Regime*. Minneapolis: University of Minnesota Press.

Ruge, Edmund. 2020. "Black self-determination drawn from our roots: An interview with Eric Jackson—Black Yield Institute, Baltimore." *Journal Periferias* 5. https://revistaperiferias.org/en/materia/black-self-determination-drawn-from-our-roots/.

Shah, Nayan. 2022. *Refusal to Eat: A Century of Prison Hunger Strikes*. Berkeley: University of California Press.

Smith, Bobby J II. 2020. "Food and the Mississippi Civil Rights Movement: Re-reading the 1962–1963 Greenwood food blockade." *Food, Culture, & Society* 23(3): 382–398.

Smoyer, Amy B., and Giza Lopes. 2017. "Hungry on the inside: Prison food as concrete and symbolic punishment in a women's prison." *Punishment & Society* 19(2): 240–255.

Soble, Leslie, Kathryn Stroud, and Marika Weinstein. 2020. "Eating behind bars: Ending the hidden punishment of food in prison. impact justice." Impact/Justice. impactjustice.org/impact/food-in-prison/#report.

Ugelvik, Thomas. 2011. "The hidden food: Mealtime resistance and identity work in a Norwegian prison." *Punishment & Society* 13(1): 47–63.

Wang, Jackie. 2018. *Carceral Capitalism*. South Pasadena, CA: Semiotext(e).

White, Monica M. 2017. *Freedom Farmers: Agricultural Resistance and the Black Freedom Movement*. Chapel Hill: University of North Carolina Press.

Winston, Celeste. 2021. "Maroon geographies." *Annals of the American Association of Geographers* 111(7): 2185–2199.

Montenegro, Maywa, and Carrie Freshour (moderators). 2021. "Abolitionist food futures" (roundtable), June 9–15. Just Food, New York University.

21 Concluding Thoughts

Alison Hope Alkon and Julian Agyeman

In *Nurturing Food Justice,* our chapter authors have shared stories of various communities' struggles for food justice on the land, within institutions, and in connection with other transformative social movements. These stories are embedded in diverse theoretical conversations seeking to make sense of the ongoing and intersecting violences of racism, settler-colonialism, heteropatriarchy, capitalism, and environmental degradation while envisioning new ways of seeing and being. Food justice also has much to contribute to our current political moment: pushing for agroecological cultivation in the context of climate collapse, offering land-based forms of healing to those on the frontlines of systemic violence, increasing food access among people experiencing poverty, and improving the lives and livelihoods of workers across the food chain.

We organized this volume according to perspective. "Dreams" shows us how, for BIPOC and queer communities, growing and serving ancestral and desired foods becomes a way to build new worlds amid the everyday violences they endure, a contemporary parallel, as chapter 4 explores, to the plots that provided a morsel of autonomy for enslaved Africans. The stories these prefigurative projects craft are in dialogue with radical theoretical traditions, from Black geographies and ecologies to Indigenous resurgence to queer epistemologies, providing embodied and grounded examples of praxis.

While queer and BIPOC food practices lie at the heart of the food justice movement, activists ignore the resurgent forces of white supremacy, uneven development, and hegemonic understandings of "good food" at their peril. Our second section, "Confrontations" looks outward to the systems in which queer and BIPOC food justice activists must operate: the racialized production of food apartheid, medicalized and colonial notions of health, and the continued privileging of white (and white supremacist) perspectives within nominally progressive food movement spaces. While some food justice activists cultivate autonomous projects that can build new worlds, others carve out spaces through "Encounters" with various institutions. The third section of this volume explores how food justice activists work within and can even shift the priorities of a variety of institutions such as food assistance programs, urban planning departments, universities, and the technosphere. The volume ends by asking how food justice has and can continue to contribute to kindred social movements for worker's rights, climate justice, and prison abolition, highlighting the multiple ways that all of these activists work to create just and sustainable socioecological relationships.

And yet, there are productive tensions between some of the chapters and the ecological niches within the ecosystem of food justice research that can provide nurturing habitat for future contributions. We highlight those here in our concluding thoughts in the hopes of offering fertile ground in which new seeds can be planted.

Many early food justice organizations were animated by a need to address diet-related health disparities such as heart disease, diabetes, and obesity (which is not actually a health condition but is often constructed as one). In response, these organizations advocated for greater access to fresh produce and other foods that the public health nutrition system deemed salubrious. In the years since we published *Cultivating Food Justice*, scholars have interrogated the overlapping cultural and scientific processes through which the health of foods is determined, and whose definitions are foregrounded and reified by state agencies and popular culture. Charlotte Biltekoff (2013, 2024) and Aaron Bobrow-Strain (2012) published exemplary books revealing the cultural logics through which debates about what to eat shape ideas about subjecthood and citizenship while ignoring structural constraints. Sabrina Strings (2020) and Julie Guthman (2011) showed how our ongoing moral panic about obesity is rooted in racism, sexism, and capitalism. Allison and Jessica Hayes-Conroy's edited volume *Doing Nutrition Differently* (2013) linked these critical analyses directly to food justice praxis through a series of cases in which communities struggle to improve community health outcomes by drawing on and imagining foodways that counter hegemonic nutrition. In this volume, Lucy Aphramor (chapter 9) takes up this debate by interrogating the public health community's colonial notion of health and working to replace it with a liberatory approach to collective flourishing. Lex Wiley and

Akilah Chatman (chapter 4) offer an alternative to debates about nutrition in food justice by describing the produce they cultivate as heritage foods rather than healthy ones. But the tension between questioning health and providing "healthy" foods remains relevant to food justice praxis. In antiracist movements, individualistic attention to diets and even dietary disparities are increasingly contextualized in understandings of medical racism and access to health care more broadly. Most food justice activists are dedicated to meeting communities "where they are at" while also working to shift individual eating practices. We encourage scholars to learn from and support these activists as their work responds to evolving understandings of health and with careful attention to the ways that individual-focused discourses of healthism can work against broader food justice goals.

Another tension that we highlighted in *Cultivating Food Justice* and that remains important today sits between communities' struggles for autonomy and desires to transform the industrial food system. The gardens cultivated by Wiley and Chatman (chapter 4) and described by Ramírez (chapter 8) offer sustenance to their communities as well as spaces to prioritize heritage crops and cultural foodways. And they are fertile grounds from which to imagine the collective flourishing of diverse identities and traditions. But such spaces remain vulnerable to the forces of racial capitalism, and Ramírez's chapter shows how the desires of city officials and white urban farmers can disrupt the good work these gardens have done. By highlighting the essential nature of land to any struggle for food justice, Ramírez hints at the potential limitations of prefigurative approaches; they build their worlds within the contemporary limits of racial capitalism and settler-colonialism while working to envision a different future. But at the same time, their abilities to exist are often threatened by these same forces. And even absent this threat, they exist alongside exploitative forces rather than confronting them.

In contrast, the chapters that look to transform some aspect of the food system—worker's rights, urban food planning, the carceral state or the ideologies of the public health nutrition system—tend to move further from the land-based agroecological practices that root these alternative projects. Hannah Garth (2020) depicts prefigurative food justice projects as more radical because they ask participants to remake their ways of seeing and being on a deeper level while characterizing broader policy-based struggles as also important but reformist. We want to push back against this hierarchy. Instead, we want to encourage future scholarship that further explores the relationships between these approaches. We have much to learn about how agroecological practices can make inroads into institutions that reach people experiencing food insecurity and how such institutions might help strengthen the practices of farmers and gardeners, especially BIPOC producers. The food justice literature remains predominantly rooted in the garden, the community, and the individual organization. While this is essential, we also want to follow its rhizomes as they spread.

We also want to note this volume's lack of research on domestic land reform efforts, including the settlements from the Pigford, Keepseagle, Garcia, and Love collective action lawsuits through which Black, Indigenous, Latinx, and Women farmers and their decedents sued the USDA for unjust treatment (Jett 2021; Figueroa and Penniman 2020). Collectives such as the Federation of Southern Cooperatives and the Disparity to Parity project have built on these settlements in order to access land and other support for BIPOC and women farmers (White 2019; Graddy-Lovelace and Naylor 2021). Organizations such as these have made significant inroads into federal policy; the USDA's socially disadvantaged farmer program and community food project grants offer funding and services to both BIPOC producers and food justice organizations (Asare-Baah et al. 2018), and groups such as the HEAL Food Collaborative, the Federation of Southern Cooperatives, and American University's Antiracist Research & Policy Center are working to amplify food and racial justice concerns in the farm bill. We would love to read critical analyses of how these opportunities shape the social, economic, and ecological imaginaries and realities of activists and farmers. At the same time, we would remind scholars following this line of inquiry, that while the work of many campaigns is to make the state respond, achievable policy proposals risk becoming "advertisements for the state" and its ability to address social problems absent collective action (Thompson and Watson 2016). Food justice movements working with, through, and outside the state can instead work to nurture what abolitionists call "nonreformist reforms" that make meaningful immediate differences in people's lives while simultaneously pushing us to demand more (Gorz 1987; Gilmore 2007; Hayes and Kaba 2023). With regard to food and agriculture, Julie Guthman's most recent book offers a parallel program of "response" rather than a narrow focus on immediate solutions (2024).

Another tension within the food justice movement and its attendant literature exists between its anticapitalist or capitalist-critical ideologies and the economic needs of the businesses that comprise it. Antonio Roman-Alcalá's extensive interviews with BIPOC food and farming activists in chapter 1 make it clear that food justice activists understand the exploitative nature of capitalist agribusiness while struggling with their own abilities to make a living within that system, whether by selling produce or receiving grants. While we agree with Roman-Alcalá that some of the critiques of neoliberal reproduction within food justice organizations have become dated, and apply differently or not at all to those informed by radical BIPOC traditions, we encourage future scholars not to abandon a critical investigation of how racial capitalism constrains the movement. And, moreover, we encourage activists to remain focused on transforming capitalist agribusiness rather than providing what Julie Guthman (2008) called "alternatives to the alternatives" of the white-dominated food movement. Ash McLeod and Eleanor Ouimet's thick description of acclaimed Indigenous-owned Minneapolis restaurant Owamni is instructive, as they seek to understand how employees navigate tensions between the restaurant's need to earn profits and its decolonial goals. This mode of analysis could also be extended to the numerous food hubs, commercial kitchens, and cooperative grocery stores that food justice activists have created in order to meet their communities' needs.

Absent this focus on capitalism, food justice risks settling for a more diverse class of elite foodies. In their now-classic book, Johnston and Baumann (2009) describe how the consumption of varied foods from diverse cultures acts as a form of social distinction for the affluent. When the book was published, the overwhelming majority of award-winning chefs were white and cooking within European food traditions. Johnson and Baumann described foodie consumers as democratized due to their desire to also explore "hole in the wall" ethnic eateries, but the low cost of these institutions served as a measure of authenticity (see also Ray 2016). Today's gastronomic scene is far more diverse, with BIPOC chefs winning awards (including Owamni's Sean Sherman), becoming notable food writers, and hosting and appearing in food-based travel shows. Representation clearly matters, and many of these BIPOC foodies bring with them not only culinary skill and vision but detailed attention to food history and a deep cultural analysis that uses food to rethink our collective past and present. In that vein, this assemblage of BIPOC elite food professionals do much to push us toward increased diversity, inclusion, and belonging. But a more racially integrated yet economically stratified capitalist food system falls far short of the capacious dreams of just food futures put forward by the authors in this volume. Indeed, increasing food access is among many food justice organizations' primary objectives, and a more diverse set of elite foodways cannot achieve this goal.

Even as the food justice movement was being nurtured by the dreams and practices of activists, capitalist agribusiness has increased its hold on global food systems. Food processing and retail companies are more consolidated than ever and are earning record profits despite the supply chain disruptions wrought by COVID-19, climate change, and Russia's war in Ukraine (Cronin 2023). According to a 2022 report by the Economic Policy Institute, these profits, and not the supply chain disruptions, are primarily responsible for inflated prices (Bivens 2022). Meanwhile, policies designed to mitigate food insecurity during the pandemic, such as the expanded child tax credit, were allowed to expire, leading to the largest recorded rise in the poverty rate in the past fifty years (Parrott 2023). Additionally, what Mansfield et al. (2023) call the "global pesticide complex" has expanded and transformed in the past two decades, resulting in both intensified use and geographic expansion. And yet, food justice scholars remain focused on activism and alternatives rather than investigating the ongoing effects of corporate agribusiness. We are not denying that movement-based scholarship is essential, but the absence of industrial agriculture in this volume is striking. There is important research on agribusiness being done in the broader world of critical food studies, including notable recent books like Madeline Fairbairn's (2020) *Fields of Gold*, which examines the financialization of global farmland, and Alex Blanchette's (2020) *Porkopolis*, which describes how the standardization practices of the pork industry transform the lives and livelihoods of its rural workers. And we are very much looking forward to chapter 20 coauthor Carrie Freshour's forthcoming book, *Making Life Work*, about the social worlds of Black women employed in Northeast Georgia's chicken industry. In addition, food justice scholars would also do well to keep abreast of potential disruptions from Silicon Valley and the agrifood tech sector (Guthman 2024; Fairbairn, Kish, and Guthman 2022). A more foundational understanding of the increasingly consolidated industrial and technologized agrifood system might help inspire dreams of the diverse agroecological food systems we want to create and the pathways through which we can get there. Food justice scholars with the time and educational privilege necessary to investigate these complex entities might serve the movement well through these sorts of analyses.

Like its predecessor, *Nurturing Food Justice* has attempted to provide an overview of the food justice movement and the activist visions, desires, stories, and projects that comprise it. A product of activist scholarship, the book depicts the nurturing work that the movement can do to increase food access, envision food sovereignty, and dream new socioecological relationships into being. In dialogue with radical social theories and kindred social movements, food justice cultivates the spaces from which to seed new social worlds beyond racial capitalism, settler colonialism, heteropatriarchy, and climate collapse. In addition to this prefigurative work, the movement nurtures efforts to transform the lives and livelihoods of food workers, the food options available to those experiencing hunger, and the ways we understand and inhabit urban and rural spaces and our own bodies. To that end, activists have carved out space for their efforts within food assistance programs, hegemonic public health nutrition discourses, urban planning processes, and neoliberal universities. Moreover, the food justice movement has cross-pollinated with and has been radicalized by movements for Black lives; Indigenous sovereignty; queer, trans, immigrant, and worker rights; climate justice; and abolition. We hope this volume can orient scholars, students, and activists to the diverse and comingled garden of projects, ideas, and stories that the food justice movement has nurtured and inspire readers to undertake plantings of their own in the seasons to come.

References

Asare-Baah, Lucy, Robert Zabawa, and Henry Findlay. 2018. "Participation in Selected USDA Programs by Socially Disadvantaged Farmers in Selected Black Belt Counties in Georgia." *Journal of Rural Social Sciences*, 33 (1): 33–55.

Biltekoff, Charlotte. 2024. *Real Food, Real Facts: Processed Food and the Politics of Knowledge*. Berkeley: University of California Press.

Biltekoff, Charlotte. 2013. *Eating Right in America: The Cultural Politics of Food and Health*. Berkeley: University of California Press.

Bivens, Josh. 2022. "Corporate Profits Have Contributed Disproportionately to Inflation. How Should Policymakers Respond?" Economic Policy Institute. https://www.epi.org/blog/corporate-profits-have-contributed-disproportionately-to-inflation-how-should-policymakers-respond/.

Blanchette, Alex. 2020. *Porkopolis*. Durham, NC: Duke University Press.

brown, adrienne marie. 2017. *Emergent Strategy*. Chico, CA: AK Press.

Cronin, Dana. 2023. "Food Prices Are Still High. What Role Do Corporate Profits Play?" Civil Eats. https://civileats.com/2023/05/22/food-prices-are-still-high-what-role-do-corporate-profits-play/.

Fairbairn, Madeline. 2020. *Fields of Gold*. Ithaca, NY: Cornell University Press.

Fairbairn, Madeline, Zenia Kish, and Julie Guthman. 2022. "Pitching Agri-Food Tech: Performativity and Non-Disruptive Disruption in Silicon Valley." *Journal of Cultural Economy* 15 (5): 652–670. https://doi.org/10.1080/17530350.2022.2085142.

Figueroa, Meleiza, and Leah Penniman. 2017. "Land Access for Beginning and Disadvantaged Farmers." *Green New Deal Policy Series: Food and Agriculture*. https://www.filesforprogress.org/memos/land_access_for_beginning_disadvantaged_farmers.pdf.

Garth, Hanna. 2020. "Blackness and 'Justice' in the Los Angeles Food Justice Movement." In *Black Food Matters*, edited by Hanna Garth and Ashanté M. Reese. Minneapolis: University of Minnesota Press.

Gorz, André. 1987. *Strategy for Labor: A Radical Proposal*. Boston: Beacon Press.

Graddy-Lovelace, Garrett, and Patti Naylor. 2021. "Disparity to Parity to Solidarity: Balancing the Scales of International Agricultural Policy for Justice and Viability." *Development* 64: 259–265.

Guthman, Julie. 2024. *The Problem with Solutions: Why Silicon Valley Can't Hack the Future of Foods*. Berkeley: University of California Press.

Guthman, Julie. 2011. "'If They Only Knew': The Unbearable Whiteness of Alternative Food." In *Cultivating Food Justice*, edited by Alison Hope Alkon and Julian Agyeman, 263–282. Cambridge, MA: MIT Press.

Hayes, Kelly, and Mariame Kaba. 2023. *Let This Radicalize You*. Chicago: Haymarket Books.

Jett, Terri. 2021. *Fighting for Farming Justice*. New York: Routledge.

Johnston, Joseé, and Shylon Baumann. 2009. *Foodies*. New York: Routledge.

Mansfield, Becky et al. 2023. "A New Critical Social Science Research Agenda on Pesticides." *Agriculture and Human Values* 41(2): 395–412.

Parrott, Sharon. 2023. "Record Rise in Poverty Highlights Importance of Child Tax Credit." Center on Budget and Policy Priorities. https://www.cbpp.org/press/statements/record-rise-in-poverty-highlights-importance-of-child-tax-credit-health-coverage.

Ray, Krishnendu. 2016. *The Ethnic Restauranteur*. New York: Bloomsbury.

Thompson, Becky, and Veronica Watson. 2018. "Theorizing White Racial Trauma and Its Remedies." In *The Construction of Whiteness*, edited by Stephen Middleton, D. Roediger, and Donald Shaffer.

Index

Publisher contact:
The MIT Press
Massachusetts Institute of Technology
77 Massachusetts Avenue, Cambridge, MA 02139
mitpress.mit.edu

EU Authorised Representative:
Easy Access System Europe, Mustamäe tee 50,
10621 Tallinn, Estonia
gpsr.requests@easproject.com

Printed by Integrated Books International,
United States of America